T0177434

Oberwolfach Seminars
Volume 43

David J. Benson
Srikanth Iyengar
Henning Krause

Representations of
Finite Groups:
Local Cohomology
and Support

 Birkhäuser

David J. Benson
Institute of Mathematics
University of Aberdeen
Fraser Noble Building
King's College
Aberdeen AB24 3UE
Scotland, UK

Srikanth Iyengar
Department of Mathematics
University of Nebraska
Lincoln, NE 68588-0130
USA

Henning Krause
Fakultät für Mathematik
Universität Bielefeld
P.O. Box 10 01 31
33501 Bielefeld
Germany

ISBN 978-3-0348-0259-8   e-ISBN 978-3-0348-0260-4
DOI 10.1007/978-3-0348-0260-4
Springer Basel Dordrecht Heidelberg London New York

Library of Congress Control Number: 2011941496

Mathematics Subject Classification (2010): 20J06, 13D99, 16E45

Printed on acid-free paper

Springer Basel AG is part of Springer Science + Business Media
(www.birkhauser-science.com)

Contents

Preface

These are the notes from an Oberwolfach Seminar which we ran from 23–29 May 2010. There were 24 graduate student and postdoctoral participants. Each morning consisted of three lectures, one from each of the organisers. The afternoons consisted of problem sessions, apart from Wednesday which was reserved for the traditional hike to St. Roman. We have tried to be reasonably faithful to the lectures and problem sessions in these notes, and have added only a small amount of new material for clarification.

The seminar focused on recent developments in classification methods in commutative algebra, group representation theory and algebraic topology. These methods were initiated by Hopkins back in 1987 [35], with the classification of the thick subcategories of the derived category of bounded complexes of finitely generated projective modules over a commutative noetherian ring R, in terms of specialisation closed subsets of $\operatorname{Spec} R$. Neeman [44] (1992) clarified Hopkins' theorem and used analogous methods to classify the localising subcategories of the derived category of unbounded complexes of modules in terms of arbitrary subsets of $\operatorname{Spec} R$. In 1997, Benson, Carlson and Rickard [9] proved the thick subcategory theorem for modular representation theory of a finite p-group G over an algebraically closed field k of characteristic p. Namely, the thick subcategories of the stable category of finitely generated kG-modules are classified by the specialisation closed subsets of the homogeneous non-maximal prime ideals in $H^*(G, k)$, the cohomology ring. The corresponding theorem for the localising subcategories of the stable category of all kG-modules has only recently been achieved, in the paper [11] by the three organisers of the seminar.

	Thick subcategories of compact objects	Localising subcategories of all objects
$\mathsf{D}(R)$	Hopkins 1987	Neeman 1992
$\mathsf{StMod}(kG)$	Benson, Carlson and Rickard 1997	Benson, Iyengar and Krause 2008

In the process of achieving the classification of the localising subcategories of StMod(kG), a general machinery was established for such classification theorems in a triangulated category; see [10, 12]. It is also worth mentioning at this stage the work of Hovey, Palmieri and Strickland [36], who did a great deal to clarify the appropriate settings for these theorems.

The general setup involves a graded commutative noetherian ring R acting on a compactly generated triangulated category with small coproducts T. Write Spec R for the set of homogeneous prime ideals of R. For each $\mathfrak{p} \in$ Spec R there is a *local cohomology functor* $\Gamma_\mathfrak{p} : T \to T$. The *support* of an object X is defined to be the subset of Spec R consisting of those $\mathfrak{p}$ such that $\Gamma_\mathfrak{p} X$ is non-zero.

The object of the game is to establish conditions under which this notion of support classifies the localising subcategories of T. This is given in terms of two conditions. The first is the *local-global principle* that says for each object X in T, the localising subcategory of T generated by X is the same as that generated by $\{\Gamma_\mathfrak{p} X\}$ as $\mathfrak{p}$ runs over the elements of Spec R. The second is a minimality condition, which requires that each $\Gamma_\mathfrak{p} T$ is either a minimal localising subcategory of T or it is zero. Under these two conditions, we say that T is *stratified* by the action of R, and then we obtain a classification theorem.

In the case of the derived category D(R), Neeman's classification made essential use of the existence of "field objects" – for a prime ideal $\mathfrak{p}$ of R, the field object is the complex consisting of the field of fractions of $R/\mathfrak{p}$, concentrated in a single degree. One of the principle obstructions to carrying out the classification in the finite group case is a lack of field objects; the obstruction theory of Benson, Krause and Schwede [15, 16] can be used to show that the required field objects usually do not exist. Circumventing this involves an elaborate series of changes of category, and machinery for transferring stratification along such changes of category. For a general finite group, the strategy is first to use Quillen stratification to reduce to elementary abelian p-groups, where there are still not enough field objects, but then to use a Koszul construction to reduce to an exterior algebra for which there are enough field objects. At this stage, a version of the Bernstein–Gelfand–Gelfand correspondence can be used to get to a graded polynomial ring, where the problem is solved. One consequence of this strategy is that we obtain classification theorems in a number of situations along the route.

In these notes we manage to give a complete proof in the case of characteristic two, where matters are considerably simplified by the fact that the group algebra of an elementary abelian 2-group is already an exterior algebra. We found it frustrating that in spite of having an entire week of lectures to explain the theory, we were not able to give a complete proof of the classification theorem for localising subcategories of StMod(kG), in odd characteristic. An overview of the classification in general characteristic is given in Section 3.3, while the proof in characteristic two may be obtained by combining Theorems 5.4 and 5.19 with results from Section 3.3.

A guide to these notes

In this volume, we have attempted to stick as closely as possible to the format of the Oberwolfach seminar. So the notes are divided into five chapters with four sections in each, corresponding to the five days with three lectures each morning and a problem session in the afternoon. The lecturing, and writing, styles of the three authors are different, and we have not tried to alter that for the purpose of these notes. In particular, there is a small amount of repetition. But we have tried to be consistent about important details such as notation, and grading everything cohomologically rather than homologically.

Prerequisites for this seminar consist of a solid background in algebra, including the basic theory of rings and modules, Artin–Wedderburn theory, Krull–Remak–Schmidt theorem; basic commutative algebra from the first chapters of the book of Atiyah and MacDonald; and basic homological algebra including derived functors, Ext and Tor. The appendix, describing the theory of support for modules over a commutative ring, is also necessary background material from commutative algebra that is not easy to find in the literature in the exact form in which we require it. The following books may be helpful.

[1] M. Atiyah and I. MacDonald, Commutative Algebra. Addison-Wesley, 1969.

[2] D. J. Benson, Representations and cohomology of finite groups I, II, Cambridge Studies in Advanced Mathematics 30, 31. Cambridge University Press, 2nd edition, 1998.

[3] W. Bruns and J. Herzog, Cohen–Macaulay rings, Cambridge Studies in Advanced Mathematics 39. Cambridge University Press, 2nd edition, 1998.

[4] R. Hartshorne, Local cohomology: A seminar given by A. Grothendieck (Harvard, 1961), Lecture Notes in Math. 41. Springer-Verlag, 1967.

[5] A. Neeman, Triangulated categories, Annals of Mathematics Studies 148. Princeton University Press, 2001.

[6] C. Weibel, Homological algebra, Cambridge Studies in Advanced Mathematics 38. Cambridge University Press, 1994.

About the exercises: These are from the problem sessions conducted during the seminar, though we have added a few more. Some are routine verifications/computations that have been omitted in the text, while others are quite substantial, and given with the implicit assumption (or hope) that, if necessary, readers would hunt for solutions in other sources.

Acknowledgments

The first and second authors would like to thank the Humboldt Foundation for their generous support of the research that led to the seminar. The second author was also partly supported by NSF grant DMS 0903493. The three of us thank the

Mathematisches Forschungsinstitut Oberwolfach for hosting this meeting, and the participating students for making it a lively seminar; their names are listed further below. Last, but not the least, we thank Jesse Burke, Kosmas Diveris, Claudia Köhler, Greg Stevenson and Matteo Varbaro for their comments and suggestions on a preliminary version of these notes.

Participants

Beck, Kristen A. (Arlington, U.S.A.)
Burke, Jesse (Lincoln, U.S.A.)
Chen, Xiao-Wu (Paderborn, Germany)
Diveris, Kosmas (Syracuse, U.S.A.)
Eghbali, Majid (Halle, Germany)
Henrich, Thilo (Bonn, Germany)
Hermann, Reiner (Bielefeld, Germany)
Köhler, Claudia (Paderborn, Germany)
Langer, Martin (Bonn, Germany)
Lassueur, Caroline L. (Lausanne, Switzerland)
Livesey, Michael (Aberdeen, Great Britain)
McKemey, Robert (Manchester, Great Britain)
Park, Sejong (Seoul, Korea)
Psaroudakis, Chrysostomos (Ioannina, Greece)
Purin, Marju (Syracuse, U.S.A.)
Reid, Fergus (Aberdeen, Great Britain)
Robertson, Marcy (Chicago, U.S.A.)
Sane, Sarang (Mumbai, India)
Scherotzke, Sarah (Paris, France)
Shamir, Shoham (Bergen, Norway)
Varbaro, Matteo (Genova, Italy)
Warkentin, Matthias (Chemnitz, Germany)
Witt, Emily E. (Ann Arbor, U.S.A.)
Xu, Fei (Bellaterra, Spain)

1 Monday

The first section gives a rapid introduction to the subject matter of the seminar. In particular, readers will encounter many notions and constructions, which will be defined and developed only in later sections. It is not expected that everything is to be understood in the first reading. The second section, 1.2, is a more leisurely (or, a less rapid) discussion of some of the basic theory of group algebras, while the last one, Section 1.3, is an introduction to triangulated categories.

1.1 Overview

1.1.1 Historical Perspective

This first lecture begins with a brief historical perspective on modular representation theory of finite groups, to give a context for the main results presented in this seminar.

Representation theory of finite groups began in the nineteenth century with the work of Burnside, Frobenius, Schur and others on finite-dimensional representations over $\mathbb{R}$ and $\mathbb{C}$. In this situation, we have the following theorem; see Theorem 1.33 for a more elaborate statement, and a proof.

Theorem 1.1 (Maschke, 1899). *Every representation of a finite group G over a field k of characteristic zero (or more generally, characteristic not dividing the group order) is a direct sum of irreducible representations. Equivalently, every short exact sequence of kG-modules splits.* □

If the characteristic of k does not divide $|G|$, the group order, we talk of *ordinary representation theory*. By contrast, *modular representation theory* refers to the situation where the characteristic does divide $|G|$. In this case, the regular representation of the group algebra kG is never a direct sum of irreducible representations, because the augmentation map $kG \to k$ sending each group element to the identity is a surjective module homomorphism that does not split.

Tentative beginnings of modular representation theory were made by Dickson in the early twentieth century. But it was not until the work of Richard Brauer from the 1940s–1970s that the subject really took off the ground. Brauer introduced modular characters, defect groups, the Brauer homomorphism, decomposition numbers, and so on; this is still the basic language for modular representation theory.

For Brauer, one of the principal goals of modular representation theory was to obtain information about the structure of finite groups. This was the era where a great deal of effort was going into the classification of the finite simple groups, and Brauer's methods were an integral part of this effort.

J. A. Green, in the decades spanning the 1960s–1990s, pioneered a change of emphasis from characters to modules. He introduced such tools as the Green correspondence, vertices and sources, Green's indecomposability theorem, the representation ring, etc.

This paved the way for Jon Carlson and others, starting in the 1980s, to introduce support variety techniques, which form the basis for the methods discussed in this seminar. These techniques were very successful in modular representation theory, and soon spread to adjacent fields such as restricted Lie algebras, finite group schemes, commutative algebra, and stable homotopy theory.

1.1.2   Classification of finite-dimensional modules

In ordinary representation theory, one classifies finite-dimensional modules by their characters. Two representations are isomorphic if and only if they have the same character. There are a finite number of irreducible characters, corresponding to the simple modules, and everything else is a sum of these. In particular, a module is indecomposable if and only if it is irreducible (Maschke's theorem).

In modular representation theory, non-isomorphic modules can have the same character. The best one can say is that two modules have the same Brauer character if and only if they have the same composition factors with the same multiplicities. There are only a finite number of simple modules, corresponding to the irreducible Brauer characters, and they also correspond to the factors in the Wedderburn decomposition of the semisimple algebra $kG/J(kG)$.

The Krull–Remak–Schmidt theorem tells us that every finite-dimensional (or equivalently, finitely generated) kG-module decomposes into indecomposable factors, and the set of isomorphism classes of the factors (with multiplicities) is an invariant of the module.

Let p be the characteristic of the field k. Then there are a finite number of isomorphism classes of indecomposable kG-modules if and only if the Sylow p-subgroups of G are cyclic (D. G. Higman, 1954). In this situation, we say that kG has *finite representation type*. The classification of the indecomposable modules for a group with cyclic Sylow p-subgroup, or more generally for a block with cyclic defect, was carried out by the work of Brauer, Thompson, Green, Dade and others.

If the Sylow p-subgroups of G are non-cyclic, the indecomposables are "unclassifiable" except if $p = 2$ and the Sylow 2-subgroups are in the following list.

- dihedral: $D_{2^n} = \langle x, y \mid x^{2^{n-1}} = 1,\ y^2 = 1,\ yxy = x^{-1} \rangle$, for $n \geq 2$

- generalised quaternion: $Q_{2^n} = \langle x, y \mid x^{2^{n-1}} = 1,\ y^2 = x^{2^{n-2}},\ yxy^{-1} = x^{-1}\rangle$, for $n \geq 3$

- semidihedral: $SD_{2^n} = \langle x, y \mid x^{2^{n-1}} = 1,\ y^2 = 1,\ yxy = x^{2^{n-2}-1}\rangle$, for $n \geq 4$.

In these cases we say that kG has *tame representation type*, while in all remaining cases with non-cyclic Sylow p-subgroups, we say that kG has *wild representation type*. In fact, this trichotomy between finite, tame and wild representation type occurs in general for finite-dimensional algebras over a field, by a theorem of Drozd.

Example 1.2. If $G = \mathbb{Z}/2 \times \mathbb{Z}/2 = \langle g, h\rangle$ and k is an algebraically closed field of characteristic 2, the classification of the indecomposable kG-modules is given by the following list.

- dimension 1: just the trivial module

- dimension $2n+1$ (for $n \geq 1$): two indecomposables denoted $\Omega^n(k)$ and $\Omega^{-n}(k)$

- dimension $2n$ (for $n \geq 1$): an infinite family of modules parametrized by points $\zeta \in \mathbb{P}^1(k)$, denoted L_{ζ^n}.

For example, the infinite family of two-dimensional modules in the above classification is described as follows. If $(\lambda : \mu)$ is a point in $\mathbb{P}^1(k)$, then $M_{(\lambda:\mu)}$ is the representation given by the matrices

$$g \mapsto \begin{pmatrix} 1 & 0 \\ \lambda & 1 \end{pmatrix}, \qquad h \mapsto \begin{pmatrix} 1 & 0 \\ \mu & 1 \end{pmatrix}.$$

It is easy to check that if $(\lambda : \mu)$ and $(\lambda' : \mu')$ represent the same point in $\mathbb{P}^1(k)$ (i.e., if $\lambda\mu' = \lambda'\mu$), then the representations are isomorphic.

1.1.3 Module categories

Given that the indecomposable modules are usually unclassifiable, how do we make progress in understanding them? Are there organisational principles that we can use? Can we make less refined classifications that are still useful?

In categorical language, we study the category $\mathsf{mod}(kG)$ of finitely generated kG-modules, and the larger category $\mathsf{Mod}(kG)$ of all kG-modules. Our goal is to understand "interesting" subcategories.

The first thing we need to discuss is the projective and injective modules. Recall that P is *projective* if every epimorphism $M \to P$ splits, and I is *injective* if every monomorphism $I \to M$ splits.

Theorem 1.3. *A kG-module is projective if and only if it is injective.*

Proof. Here is a sketch of a proof. One of the exercises is to fill in the details. First one proves that kG is injective as a kG-module (this is called *self-injectivity*) by giving a kG-module isomorphism between kG and its vector space dual. Since kG

is noetherian, direct sums of injective modules are injective, so free modules are injective. Direct summands of injectives are injective, so projectives are injective.

For the converse, we show that every module M embeds in a free module. Namely, if we denote by $M{\downarrow}_{\{1\}}{\uparrow}^G$ the free module obtained by restricting to the trivial subgroup and then inducing back up to G, then there is a monomorphism $M \to M{\downarrow}_{\{1\}}{\uparrow}^G$. If M is injective, then this map splits and M is a direct summand of a free module, hence projective. $\square$

Remark 1.4. It is also true that a kG-module is projective if and only if it is flat; we shall not make use of this fact. Furthermore, if G is a p-group (p is the characteristic of k), then a kG-module is projective if and only if it is free; see Proposition 1.34.

There are only a finite number of projective indecomposables, or equivalently injective indecomposables, and for any given group these can be understood. If P is projective indecomposable, then P has a unique top composition factor and a unique bottom composition factor, and they are isomorphic simple modules S. The projective module $P = P(S)$ is determined up to isomorphism by S, and this gives a one-to-one correspondence between simples and projective indecomposables. Thus $P(S)$ is both the projective cover and the injective hull of S.

1.1.4   The stable module category

Recall that the category $\mathsf{Mod}(kG)$ has as its objects the kG-modules, as its arrows the module homomorphisms.

It is often convenient to work "modulo the projective modules", which leads to the *stable module category* $\mathsf{StMod}(kG)$. This has the same objects as $\mathsf{Mod}(kG)$, but the arrows are given by quotienting out the module homomorphisms that factor through some projective module. We write

$$\underline{\mathrm{Hom}}_{kG}(M, N) = \mathrm{Hom}_{kG}(M, N)/P\,\mathrm{Hom}_{kG}(M, N)$$

where $P\,\mathrm{Hom}_{kG}(M, N)$ denotes the linear subspace consisting of homomorphisms that factor through some projective module. Note that $M \to N$ factors through some projective module if and only if it factors through the projective cover of N ($M \to P(N) \to N$), and also if and only if it factors through the injective hull of M ($M \to I(M) \to N$). This implies in particular that $P\,\mathrm{Hom}_{kG}(M, N)$ is a linear subspace of $\mathrm{Hom}_{kG}(M, N)$.

We write $\mathsf{mod}(kG)$ and $\mathsf{stmod}(kG)$ for the full subcategories of finitely generated modules in $\mathsf{Mod}(kG)$ and $\mathsf{StMod}(kG)$ respectively. Note that by the discussion above, a homomorphism of finitely generated modules factors through a projective module if and only if it factors through a finitely generated projective module.

Warning 1.5. It is not true that if a homomorphism of kG-modules factors through a finitely generated module and it factors through a projective module, then it factors through a finitely generated projective module.

The categories $\mathsf{Mod}(kG)$, $\mathsf{mod}(kG)$ are abelian categories, but $\mathsf{StMod}(kG)$, $\mathsf{stmod}(kG)$ are not. Roughly speaking, the problem is that every homomorphism $M \to N$ of kG-modules is equivalent in $\mathsf{StMod}(kG)$ to the surjective homomorphism $M \oplus P(N) \to N$ and to the injective homomorphism $M \to I(M) \oplus N$, so we lose sight of kernels and cokernels.

Instead, $\mathsf{StMod}(kG)$ and $\mathsf{stmod}(kG)$ are examples of *triangulated categories*. This is proved in detail in the book of Happel [33], and will be discussed in Section 1.3. For now, we just mention that the distinguished triangles in $\mathsf{StMod}(kG)$ come from short exact sequences in $\mathsf{Mod}(kG)$ and the shift in $\mathsf{StMod}(kG)$ is Ω^{-1}.

Next, we discuss how one tells whether a kG-module is projective. This is done through Chouinard's theorem and Dade's lemma.

1.1.5 Chouinard's theorem

Let p be a prime number and k a field of characteristic p.

Definition 1.6. A finite group is an *elementary abelian p-group* if it is isomorphic to $(\mathbb{Z}/p)^r = \mathbb{Z}/p \times \cdots \times \mathbb{Z}/p$ for some r. The number r is called the *rank*.

Theorem 1.7 (Chouinard [26] (1976)). *A kG-module is projective if and only if its restriction to every elementary abelian subgroup of G is projective.* □

Remark 1.8. In fact, Chouinard proved this theorem in the context where k is an arbitrary commutative rings of coefficients. In this case, one needs to use elementary abelian subgroups at all primes dividing the group order. We shall only make use of the case where k is a field of characteristic p, in which case we only need the elementary abelian p-subgroups.

Example 1.9. Let $G = Q_8$, the quaternions and k a field of characteristic 2. The only elementary abelian subgroup of G is its centre $Z(G) = \{1, z\}$. In this case, Chouinard's theorem states that a kG-module M is projective if and only if its restriction to $Z(G)$ is projective. If the module is finite-dimensional, this is equivalent to the statement that the rank of the matrix representing $1 + z$ is as large as it can be, namely one half of the dimension of M. This can be seen using the theory of Jordan canonical forms.

1.1.6 Dade's lemma

Let E be an elementary abelian p-group of rank r:

$$E = (\mathbb{Z}/p)^r = \langle g_1, \ldots, g_r \rangle$$

and let k be an algebraically closed field of characteristic p.

Write x_i for the element $g_i - 1$ in $J(kE)$, the Jacobson radical of kE, so that

$$kE = k[x_1, \ldots, x_r]/(x_1^p, \ldots, x_r^p).$$

Definition 1.10. If $\alpha = (\alpha_1, \ldots, \alpha_r) \in \mathbb{A}^r(k) \setminus \{0\}$, we set

$$x_\alpha = \alpha_1 x_1 + \cdots + \alpha_r x_r \in J(kE)$$

so that $x_\alpha^p = 0$, and $1 + x_\alpha$ is a unit of order p in kE. We call the subgroup $\langle 1 + x_\alpha \rangle$ of the group of units of kE generated by $1 + x_\alpha$ a *cyclic shifted subgroup* of E.

Recall that projective kE-modules (equivalently, injective modules) are free.

Lemma 1.11 (Dade [27] (1978)). *A finitely generated kE-module M is projective if and only if its restriction to every cyclic shifted subgroup $\langle 1 + x_\alpha \rangle$ is free.* □

Remark 1.12. Using the theory of Jordan canonical forms, we can see that the restriction of M to $\langle 1 + x_\alpha \rangle$ is free if and only if the rank of the matrix representing x_α is as large as it can be, namely $\left(\frac{p-1}{p} \right) \cdot \dim_k(M)$.

1.1.7   Rank varieties

Although it is logically not necessary to discuss rank varieties for the purpose of understanding the proofs of the main results presented in this seminar, cohomological varieties, which are needed, are very difficult to compute without reference to rank varieties. We therefore include a discussion of rank varieties, first for finitely generated modules and then for infinitely generated modules.

Dade's lemma motivates the following definition.

Definition 1.13 (Carlson [23, 24]). Let E be an elementary abelian p-group and k a field of characteristic p. If M is a finitely generated kE-module, then the *rank variety* of M is

$$V_E^r(M) = \{0\} \cup \{\alpha \neq 0 \mid M{\downarrow}_{\langle 1 + x_\alpha \rangle} \text{ is not free}\} \subseteq V_E^r = \mathbb{A}^r(k)$$

Here are some properties of rank varieties; some of these are obvious, while others are difficult to prove. A detailed account of the theory can be found in Chapter 5 of [5].

(1) $V_E^r(M)$ is a closed homogeneous subvariety of $\mathbb{A}^r(k)$.

(2) $V_E^r(M) = \{0\}$ if and only if M is projective.

(3) $V_E^r(M \oplus N) = V_E^r(M) \cup V_E^r(N)$.

(4) $V_E^r(M \otimes_k N) = V_E^r(M) \cap V_E^r(N)$.

(5) In any exact sequence $0 \to M_1 \to M_2 \to M_3 \to 0$ of kE-modules the variety of each module is contained in the union of the varieties of the other two.

(6) The Krull dimension of $V_E^r(M)$ measures the rate of growth of the minimal projective resolution of M.

(7) Given a closed homogeneous subvariety V of $\mathbb{A}^r(k)$, there exists a finitely generated kE-module M such that $V_E^r(M) = V$.

1.1.8 Infinitely generated modules

Why study infinitely generated modules?

Here is an analogy. If we were only interested in finite CW complexes, or even only interested in manifolds, we would end up looking at homology theories (ordinary homology, K-theory, etc.). The representing objects for these are infinite CW complexes such as $K(\pi, n)$, $BU(n)$, etc.

Similarly, in representation theory the representing objects for some natural functors on finitely generated modules are infinitely generated. Relevant examples for us are Rickard's idempotent modules and idempotent functors [53], which we shall be discussing later.

Another relevant motivation comes from commutative algebra. Over a commutative noetherian ring, the injective hull of a finitely generated module is usually not finitely generated. Since the definition of local cohomology functors in this context involves injective resolutions, we should expect to have to deal with infinitely generated modules.

More care is needed in dealing with infinitely generated modules, as many of the well known properties of finitely generated modules fail. So for example a non-zero kG-module need not have any indecomposable direct summands. Even if it is a finite direct sum of indecomposables, the Krull–Remak–Schmidt theorem need not hold.

Warning 1.14. Dade's lemma as stated earlier in this lecture is false for infinitely generated kG-modules, as is explained in the following example.

Example 1.15. This is the *generic module* for $(\mathbb{Z}/2)^r = \langle g_1, \ldots, g_r \rangle$, for $r \geq 2$.

Let $K = k(t_1, \ldots, t_r)$, a pure transcendental extension of k of transcendence degree r, and let $M = K \oplus K$ as a k-vector space, with each g_i acting on M as the matrix

$$\begin{pmatrix} I & 0 \\ t_i & I \end{pmatrix}.$$

Here, I is the identity map on K, and t_i is to be thought of as multiplication by t_i on K. The reader is asked to prove in the exercises that M is projective on all cyclic shifted subgroups, but it is not projective.

In fact, in a sense that we are about to explain, the variety of this module M is *the generic point* of $\mathbb{A}^r(k)$.

The following modification of Dade's lemma was proved by Benson, Carlson and Rickard [8].

Lemma 1.16. *A kE-module M is projective if and only if for all extension fields K of k and all cyclic shifted subgroups $\langle 1 + x_\alpha \rangle$ of KE the module $(K \otimes_k M){\downarrow}_{\langle 1+x_\alpha \rangle}$ is free.* □

If k is algebraically closed, write $\mathcal{V}_E^r(k)$ for the set of non-zero homogeneous irreducible closed subvarieties V of $\mathbb{A}^r(k)$. Then each cyclic shifted subgroup of KE is *generic* for some $V \in \mathcal{V}_E^r(k)$ in a sense described in [8].

Definition 1.17. If M is a, possibly infinite-dimensional, kE-module, then the *rank variety* is no longer a single variety, but a set of varieties:

$$\mathcal{V}_E^r(M) = \left\{ V \in \mathcal{V}_E^r(k) \,\middle|\, \begin{array}{l} K \otimes_k M \text{ restricted to a generic cyclic} \\ \text{shifted subgroup for } V \text{ is not projective.} \end{array} \right\}$$

The following is a list of properties of the rank varieties $\mathcal{V}_E^r(M)$; cf. §1.1.7. Again, some are obvious and some are difficult to prove; see [8].

(1) $\mathcal{V}_E^r(M) = \varnothing$ if and only if M is projective.

(2) $\mathcal{V}_E^r(M \oplus N) = \mathcal{V}_E^r(M) \cup \mathcal{V}_E^r(N)$, and more generally

$$\mathcal{V}_E^r\left(\bigoplus_\alpha M_\alpha\right) = \bigcup_\alpha \mathcal{V}_E^r(M_\alpha).$$

(3) $\mathcal{V}_E^r(M \otimes_k N) = \mathcal{V}_E^r(M) \cap \mathcal{V}_E^r(N)$.

(4) If $0 \to M_1 \to M_2 \to M_3 \to 0$ is a short exact sequence of kE-modules, then the variety of each module is contained in the union of the varieties of the other two.

(5) For a finitely generated kE-module M, $\mathcal{V}_E^r(M)$ is the set of closed homogeneous irreducible subvarieties of $V_E^r(M)$.

(6) For any subset $\mathcal{V} \subseteq \mathcal{V}_E^r(k)$ there exists a kE-module M such that $\mathcal{V}_E^r(M) = \mathcal{V}$.

1.1.9 Localising subcategories

Our goal in these lectures is to understand the subcategories of $\mathsf{Mod}(kG)$ in terms of varieties. Let us state the main theorem for an elementary abelian group E.

Definition 1.18. Consider full subcategories C of $\mathsf{Mod}(kE)$ satisfying the following two properties:

(1) C is closed under set-indexed direct sums, and

(2) C has the "two in three property":

If $0 \to M_1 \to M_2 \to M_3 \to 0$ is an exact sequence of kE-modules and two of M_1, M_2, M_3 are in C, then so is the third.

Such subcategories, if they are non-zero, contain the projective modules (Exercise!) and pass down to the *localising subcategories* of $\mathsf{StMod}(kE)$.

The following is the main theorem of [11], reinterpreted for rank varieties in the case of an elementary abelian p-group.

Theorem 1.19. *The non-zero subcategories of* $\mathsf{Mod}(kE)$ *satisfying the above two conditions are in bijection with subsets of the set* $\mathcal{V}_E^r(k)$ *of non-zero closed homogeneous irreducible subvarieties of* $\mathbb{A}^r(k)$. *Under this bijection, a subset* $\mathcal{V} \subseteq \mathcal{V}_E^r(k)$ *corresponds to the full subcategory consisting of those kE-modules M satisfying* $\mathcal{V}_E^r(M) \subseteq \mathcal{V}$. $\qquad\square$

1.1.10 Thick Subcategories

The corresponding result for the finitely generated module category $\mathsf{mod}(kE)$ was proved in [9], and goes as follows.

Definition 1.20. A subset $\mathcal{V}$ of a set of varieties is *specialisation closed* if whenever $V \in \mathcal{V}$ and $W \subseteq V$ we have $W \in \mathcal{V}$.

Definition 1.21. Consider full subcategories C of $\mathsf{mod}(kE)$ satisfying the following two properties:

(1) C is closed under finite direct sums and summands, and

(2) C has the "two in three property" as before.

Such subcategories if they are non-zero, contain the projective modules and pass down to the *thick subcategories* of $\mathsf{stmod}(kE)$.

Theorem 1.22. *The non-zero subcategories of* $\mathsf{mod}(kE)$ *satisfying the above two conditions are in bijection with subsets of* $\mathcal{V}_E^r(k)$ *that are closed under specialisation. Under this bijection, a specialisation closed subset* $\mathcal{V} \subseteq \mathcal{V}_E^r(k)$ *corresponds to the full subcategory consisting of those finitely generated kE-modules M with the property that every irreducible component of $V_E^r(M)$ is an element of $\mathcal{V}$.* $\square$

1.2 Modules over group algebras

This section begins afresh, as it were, and introduces basic concepts and constructions concerning modules over group algebra. Our basic reference for this material is [4]; the commutative algebraists among readers may prefer to look also at [37].

In what follows G denotes a finite group and k a field. We write char k for the characteristic of k. The *group algebra* of G over k is the k-vector space

$$kG = \bigoplus_{g \in G} kg$$

with product induced from G. The identity element of the group, denoted by 1, is also the identity for kG and k is identified with the subring $k1$ of kG. It is a central subring, so kG is a k-algebra. Evidently, the ring kG is commutative if and only if the group G is abelian; see also Exercise 3 at the end of this chapter.

This construction is functorial, on the category of groups.

Example 1.23. When $G = \langle g \mid g^d = 1 \rangle$, a cyclic group $\mathbb{Z}/d$ of order d, one has $kG = k[x]/(x^d - 1)$, the polynomial ring in the variable x modulo the ideal generated by $x^d - 1$.

More generally, if $G = \langle g_1, \ldots, g_r \mid g_1^{d_1} = 1, \ldots, g_r^{d_r} = 1 \rangle$ is a product of cyclic groups of order $d_i \geq 1$, then

$$kG \cong k[x_1, \ldots, x_r]/(x_1^{d_1} - 1, \ldots, x_r^{d_r} - 1),$$

where x_i corresponds to the element g_i.

A further specialisation plays an important role in these lectures: For p a prime number, the abelian group $(\mathbb{Z}/p)^r$ is an *elementary abelian p-group of rank r*. When $\operatorname{char} k = p$, setting $z_i = x_i - 1$, gives an isomorphism of k-algebras

$$kG \cong k[z_1, \ldots, z_r]/(z_1^p, \ldots, z_r^p).$$

Remark 1.24. As a k-vector space, the rank of kG equals $|G|$, the order of the group G; in particular, kG is a finite-dimensional algebra over k, so it is artinian and noetherian, both on the left and on the right.

Now we move on to module theory over kG. From this perspective, there are many remarkable features that distinguish kG from arbitrary (even finite-dimensional) k-algebras. For a start one has:

Remark 1.25. Each right kG-module M is, canonically, a left module with product defined by $gm = mg^{-1}$, for $m \in M$ and $g \in G$. Said otherwise, the map $g \mapsto g^{-1}$ gives an isomorphism between kG and its opposite ring. Thus, the category of right modules is equivalent to the category of left modules.

For this reason, henceforth we focus on left modules.

Definition 1.26. Let M and N be (left) kG-modules. There is a *diagonal action* of kG on the k-vector space $M \otimes_k N$ defined by

$$g(m \otimes n) = gm \otimes gn \quad \text{for all } g \in G, \ m \in M \text{ and } n \in N.$$

One can verify that the diagonal action on $M \otimes_k N$ defines a kG-module structure as follows: Since M, N are kG-modules, $M \otimes_k N$ is a module over $kG \otimes_k kG$, with

$$(g \otimes h) \cdot (m \otimes n) = (gm \otimes hn).$$

By functoriality of the construction of group algebras, the diagonal homomorphism of groups $G \to G \times G$, where $g \mapsto (g,g)$, induces a homomorphism of k-algebras $kG \to kG \otimes_k kG$; see Exercise 1. Restriction of the $kG \otimes_k kG$-module action along this map gives the diagonal action of kG on $M \otimes_k N$.

In what follows $M \otimes_k N$ will always denote this kG-module. This is not to be confused with the action of kG on $M \otimes_k N$ induced by M (in which case, one gets a direct sum of $\dim_k N$ copies of M), or the one induced by N. These are also important and arise naturally; see Definition 1.31.

We note that there is a kG-linear isomorphism

$$M \otimes_k N \cong N \otimes_k M$$

defined by $m \otimes n \mapsto n \otimes m$.

In the same vein, $\operatorname{Hom}_k(M, N)$ is a kG-module with diagonal action, where for $g \in G$ and $\alpha \in \operatorname{Hom}_k(M, N)$ one has

$$(g \cdot \alpha)(m) = g\alpha(g^{-1}m) \quad \text{for } m \in M.$$

Definition 1.27. For each kG-module M the subset

$$M^G = \{m \in M \mid gm = m \text{ for all } g \in G\}$$

is a submodule, called the *invariant submodule* of M.

The *augmentation* of kG is the homomorphism of k-algebras

$$\varepsilon \colon kG \to k \quad \text{where} \quad \varepsilon\Big(\sum_{g \in G} c_g g\Big) = \sum_{g \in G} c_g \,.$$

That this is a homomorphism of k-algebras can be checked directly, or by noting that it is induced by the constant homomorphism $G \to \{1\}$ of groups. The kernel of ε is the (two-sided) ideal $\bigoplus_g k(g-1)$, where the sum runs over all $g \in G \setminus \{1\}$.

The kG-module structure on k thus obtained is called the *trivial* one. It is immediate from the description of $\mathrm{Ker}(\varepsilon)$ that there is identification

$$\mathrm{Hom}_{kG}(k, M) = M^G \,.$$

In particular, the functor defined on objects by $M \mapsto M^G$ is left exact; right exactness is equivalent to the projectivity of k as a kG-module; see also Theorem 1.33.

Remark 1.28. It is easy to verify that for any kG-modules M and N one has

$$\mathrm{Hom}_{kG}(M, N) = \mathrm{Hom}_k(M, N)^G \,.$$

Recall that kG acts on $\mathrm{Hom}_k(M, N)$ diagonally; see Definition 1.26.

For each kG-module L, the adjunction isomorphism of k-vector spaces:

$$\mathrm{Hom}_k(L, \mathrm{Hom}_k(M, N)) \cong \mathrm{Hom}_k(L \otimes_k M, N)$$

is compatible with the kG-module structures; this can (and should) be verified directly. Applying $(-)^G$ to it yields the *adjunction isomorphism*:

$$\mathrm{Hom}_{kG}(L, \mathrm{Hom}_k(M, N)) \cong \mathrm{Hom}_{kG}(L \otimes_k M, N) \,. \tag{1.29}$$

The adjunction isomorphism has the following remarkable consequence.

Proposition 1.30. *For any projective kG-module P the kG-modules $M \otimes_k P$ and $P \otimes_k M$ are projective.*

Note: The kG-modules $M \otimes_k kG$ and $kG \otimes_k M$ are even free; see Exercise 5.

Proof. The functors $\mathrm{Hom}_k(M, -)$ and $\mathrm{Hom}_{kG}(P, -)$ of kG-modules are exact and hence so is their composition $\mathrm{Hom}_{kG}(P, \mathrm{Hom}_k(M, -))$. The latter is isomorphic to $\mathrm{Hom}_{kG}(P \otimes_k M, -)$, by the adjunction isomorphism (1.29), so one deduces that the kG-module $P \otimes_k M$ is projective. It remains to note that this module is isomorphic, as a kG-module, to $M \otimes_k P$. $\qquad\square$

Definition 1.31. Let $H \leq G$ be a subgroup and $\iota \colon kH \subseteq kG$ the induced inclusion of k-algebras.

Restriction of scalars along ι endows each kG-module M with a structure of a kH-module, denoted $M{\downarrow}_H$. One thus gets a *restriction* functor

$$(-){\downarrow}_H \colon \operatorname{Mod} kG \to \operatorname{Mod} kH .$$

This functor is evidently exact. Base change along ι induces a functor

$$(-){\uparrow}^G = kG \otimes_{kH} - \colon \operatorname{Mod} kH \to \operatorname{Mod} kG .$$

This functor is called *induction*; it is also exact, because kG is a free kH-module; see Exercise 8. One can, and does, also consider the *coinduction* functor

$$\operatorname{Hom}_{kH}(kG, -) \colon \operatorname{Mod} kH \to \operatorname{Mod} kG .$$

For any kH-module L there is a natural isomorphism $\operatorname{Hom}_{kH}(kG, L) \cong L{\uparrow}^G$ of kG-modules. The version of the adjunction

$$\operatorname{Hom}_{kG}(L{\uparrow}^G, M) \cong \operatorname{Hom}_{kH}(L, M{\downarrow}_H) . \tag{1.32}$$

is called *Frobenius reciprocity*.

1.2.1   Structure of the ring kG

Next we recall some of the salient features and results about the group algebra. The starting point is a souped-up version of Maschke's Theorem.

Theorem 1.33. *The following conditions are equivalent.*

(1) *The ring kG is semi-simple.*

(2) *The trivial module k is projective.*

(3) *The functor $(-)^G$ on $\operatorname{Mod} kG$ is exact.*

(4) $\operatorname{char} k$ *does not divide* $|G|$.

Proof. The implication (1) $\implies$ (2) is clear.

(2) $\implies$ (4) Let $\varepsilon \colon kG \to k$ be the augmentation homomorphism, which defines the trivial action on k. When k is projective, there exists a homomorphism $\sigma \colon k \to kG$ of kG-modules such that $\varepsilon \circ \sigma = \operatorname{id}_k$. Write

$$\sigma(1) = \sum_{g \in G} c_g g .$$

For each $h \in G$ one has $\sigma(1) = \sigma(h^{-1} \cdot 1) = h^{-1} \sum_{g \in G} c_g g$. So comparing coefficients of the identity element in G one gets that $c_h = c_1$. Thus, $\sigma(1) = c_1 \sum_{g \in G} g$ so that in k there is an equality $1 = \varepsilon \sigma(1) = c_1 |G|$; in particular $|G| \neq 0$.

(4) $\implies$ (3) Let M be a kG-module. Consider the map

$$\rho\colon M \to M \quad \text{where } \rho(m) = \frac{1}{|G|} \sum_{g \in G} gm.$$

A straightforward computation shows that this map is kG-linear and an identity when restricted to M^G, the submodule of invariants. The construction is evidently functorial, which means that $(-)^G$ is a direct summand of the identity functor, and since the latter is exact so is the former.

(3) $\implies$ (1) It suffices to prove that any epimorphism $M \twoheadrightarrow N$ of kG-modules splits. Any such map induces an epimorphism $\mathrm{Hom}_k(N, M) \twoheadrightarrow \mathrm{Hom}_k(N, N)$, and hence, since $(-)^G$ is exact, also an epimorphism

$$\mathrm{Hom}_{kG}(N, M) = \mathrm{Hom}_k(N, M)^G \twoheadrightarrow \mathrm{Hom}_k(N, N)^G = \mathrm{Hom}_{kG}(N, N).$$

Then the identity on N lifts to a kG-linear map $N \to M$, as desired. $\qquad\square$

In any characteristic, the group algebra kG has the following properties:

- The Krull–Remak–Schmidt theorem holds in $\mathsf{mod}\, kG$.

- The group algebra kG is *self-injective*, meaning, that it is injective as a module over itself. A proof of this assertion is sketched in Theorem 1.3, and also part of Monday's exercises.

- There are canonical bijections of isomorphism classes

$$\left\{ \text{Simple modules} \right\} \leftrightarrow \left\{ \begin{array}{c} \text{indecomposable} \\ \text{projectives} \end{array} \right\} \leftrightarrow \left\{ \begin{array}{c} \text{indecomposable} \\ \text{injectives} \end{array} \right\}$$

where Artin–Wedderburn theory gives the one on the left, and the one on the right holds by self-injectivity of kG.

Given these properties, it is not hard to prove the following characterization of p-groups; for a proof, see, for instance, [37, (1.5) and (1.6)].

Proposition 1.34. *The following conditions are equivalent.*

(1) *The ring kG is local; i.e., it has a unique maximal ideal.*

(2) *The trivial module k is the only simple module.*

(3) *G is a p-group.* $\qquad\square$

1.2.2 Group cohomology

In the remainder of this section we assume $\mathrm{char}\, k$ divides $|G|$; equivalently, $(-)^G$ is not exact; see Theorem 1.33. For each kG-module M and integer n, the nth *cohomology of G with coefficients in M* is the k-vector space

$$H^n(G; M) = \mathrm{Ext}^n_{kG}(k, M).$$

We speak of $H^n(G; k)$ as the cohomology of G. Yoneda composition induces on the graded k-vector space $H^*(G; k)$ the structure of a k-algebra, and on $H^*(G; M)$ the structure of a graded right $H^*(G, k)$-module.

Recall that a graded ring R is said to be *graded commutative* if

$$r \cdot s = (-1)^{|r||s|} s \cdot r \quad \text{for all } r, s \in R.$$

Here is a remarkable feature of the cohomology algebra of a group:

Proposition 1.35. *The k-algebra $H^*(G, k)$ is graded commutative.* $\qquad\square$

The key point is that kG is in fact a Hopf algebra, with diagonal induced by the diagonal homomorphism $G \to G \times G$; see, for instance, [37, Proposition 5.5].

The cohomology algebra of an elementary abelian group is easy to compute, and has been known for a long time; see [25, Chapter XII, §7]. We give the argument for 2-groups; see Section 2.3 for a different perspective on this computation and a more detailed discussion of the group cohomology algebra.

Proposition 1.36. *Let $G = (\mathbb{Z}/2)^r$ and char $k = 2$. Then $H^*(G, k)$ is a polynomial algebra over k on r variables each of degree 1:*

$$H^*(G, k) \cong k[y_1, \ldots, y_r] \quad \text{where } |y_i| = 1.$$

Proof. Consider first the case where $r = 1$, so $G = \mathbb{Z}/2$. The claim is then that there is an isomorphism of k-algebras $H^*(G; k) \cong k[y]$, where $|y| = 1$. Indeed, the group algebra kG is then isomorphic to $k[x]/(x^2)$, and the complex

$$\cdots \to kG \xrightarrow{x} kG \xrightarrow{x} kG \to 0,$$

is a free resolution of the trivial kG-module k. It follows that $\operatorname{Ext}^n_{kG}(k, k) \cong k$ for each $n \geq 0$, and that the class of the extension

$$0 \to k \xrightarrow{\eta} kG \xrightarrow{\varepsilon} k \to 0,$$

where $\eta(1) = x$ and ε is the augmentation, generates $H^*(G, k)$ as a k-algebra.

The structure of the cohomology algebra for $r \geq 1$ then follows from repeated applications of the Künneth isomorphism:

$$\operatorname{Ext}_{A \otimes_k B}(k, k) \cong \operatorname{Ext}_A(k, k) \otimes_k \operatorname{Ext}_B(k, k)$$

where A and B are augmented k-algebras; see [25, Chapter XI]. $\qquad\square$

Here is the corresponding result for odd primes. It can be proved along the same lines as the preceding one.

Proposition 1.37. *Let p be an odd prime, $G = (\mathbb{Z}/p)^r$ and char $k = p$. Then $H^*(G, k)$ is the tensor product of an exterior algebra in r variables in degree 1 and a polynomial algebra in r variables in degree 2:*

$$H^*(G, k) \cong \left(\bigwedge \bigoplus_{i=1}^r ky_i \right) \otimes_k k[z_1, \ldots, z_r] \quad \text{where } |y_i| = 1 \text{ and } |z_i| = 2.$$

1.3 Triangulated categories

This lecture provides a quick introduction to triangulated categories. We give definitions and explain the basic concepts. The triangulated categories arising in this work are always *algebraic* which means that they are equivalent to the stable category of some Frobenius category. We do not use this fact explicitly, but it helps sometimes to understand the triangulated structure.

Triangulated categories were introduced by Verdier in his thesis which was published posthumously [54]; it is still an excellent reference. Another source for the material in this section is [42]. For the material on exact categories and Frobenius categories see Happel [33] and also Bühler's survey article [22].

1.3.1 Triangulated categories

Let T be an additive category together with a fixed equivalence $\Sigma \colon \mathsf{T} \xrightarrow{\sim} \mathsf{T}$, which one calls *shift* or *suspension*. A *triangle* in T is a sequence (α, β, γ) of morphisms

$$X \xrightarrow{\alpha} Y \xrightarrow{\beta} Z \xrightarrow{\gamma} \Sigma X ,$$

and a morphism between triangles (α, β, γ) and $(\alpha', \beta', \gamma')$ is a triple (ϕ_1, ϕ_2, ϕ_3) of morphisms in T making the following diagram commutative.

$$
\begin{array}{ccccccc}
X & \xrightarrow{\alpha} & Y & \xrightarrow{\beta} & Z & \xrightarrow{\gamma} & \Sigma X \\
\downarrow{\phi_1} & & \downarrow{\phi_2} & & \downarrow{\phi_3} & & \downarrow{\Sigma \phi_1} \\
X' & \xrightarrow{\alpha'} & Y' & \xrightarrow{\beta'} & Z' & \xrightarrow{\gamma'} & \Sigma X'
\end{array}
$$

The category T is called *triangulated* if it is equipped with a class of distinguished triangles (called *exact triangles*) satisfying the following conditions.

(T1) Any triangle isomorphic to an exact triangle is exact. For each object X, the triangle $0 \to X \xrightarrow{\mathrm{id}} X \to 0$ is exact. Each morphism α fits into an exact triangle (α, β, γ).

(T2) A triangle (α, β, γ) is exact if and only if $(\beta, \gamma, -\Sigma\alpha)$ is exact.

(T3) Given exact triangles (α, β, γ) and $(\alpha', \beta', \gamma')$, each pair of morphisms ϕ_1 and ϕ_2 satisfying $\phi_2\alpha = \alpha'\phi_1$ can be completed to a morphism of triangles:

$$
\begin{array}{ccccccc}
X & \xrightarrow{\alpha} & Y & \xrightarrow{\beta} & Z & \xrightarrow{\gamma} & \Sigma X \\
\downarrow{\phi_1} & & \downarrow{\phi_2} & & \downarrow{\phi_3} & & \downarrow{\Sigma \phi_1} \\
X' & \xrightarrow{\alpha'} & Y' & \xrightarrow{\beta'} & Z' & \xrightarrow{\gamma'} & \Sigma X'
\end{array}
$$

(T4) Given exact triangles $(\alpha_1, \alpha_2, \alpha_3)$, $(\beta_1, \beta_2, \beta_3)$, and $(\gamma_1, \gamma_2, \gamma_3)$ with $\gamma_1 = \beta_1 \alpha_1$, there exists an exact triangle $(\delta_1, \delta_2, \delta_3)$ making the following diagram commutative.

$$
\begin{array}{ccccccc}
X & \xrightarrow{\alpha_1} & Y & \xrightarrow{\alpha_2} & U & \xrightarrow{\alpha_3} & \Sigma X \\
\| & & \downarrow{\scriptstyle \beta_1} & & \downarrow{\scriptstyle \delta_1} & & \| \\
X & \xrightarrow{\gamma_1} & Z & \xrightarrow{\gamma_2} & V & \xrightarrow{\gamma_3} & \Sigma X \\
& & \downarrow{\scriptstyle \beta_2} & & \downarrow{\scriptstyle \delta_2} & & \downarrow{\scriptstyle \Sigma\alpha_1} \\
& & W & = & W & \xrightarrow{\beta_3} & \Sigma Y \\
& & \downarrow{\scriptstyle \beta_3} & & \downarrow{\scriptstyle \delta_3} & & \\
& & \Sigma Y & \xrightarrow{\Sigma\alpha_2} & \Sigma U & &
\end{array}
$$

The axiom (T4) is known as *octahedral axiom* because the four exact triangles can be arranged in a diagram having the shape of an octahedron. The exact triangles $A \to B \to C \to \Sigma A$ are represented by faces of the form

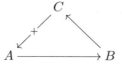

and the other four faces are commutative triangles.

Let us give a more intuitive formulation of the octahedral axiom which is based on the notion of a homotopy cartesian square. Call a commutative square

$$
\begin{array}{ccc}
X & \xrightarrow{\alpha'} & Y' \\
\downarrow{\scriptstyle \alpha''} & & \downarrow{\scriptstyle \beta'} \\
Y'' & \xrightarrow{\beta''} & Z
\end{array}
$$

homotopy cartesian if there exists an exact triangle

$$
X \xrightarrow{\left[\begin{smallmatrix} \alpha' \\ \alpha'' \end{smallmatrix}\right]} Y' \amalg Y'' \xrightarrow{[\beta' \ -\beta'']} Z \xrightarrow{\gamma} \Sigma X.
$$

The morphism γ is called a *differential* of the homotopy cartesian square. Note that a differential of the homotopy cartesian square changes its sign if the square is flipped along the main diagonal.

Assuming (T1)–(T3), one can show that (T4) is equivalent to the following condition; see [42, §2.2].

(T4′) Every pair of morphisms $X \to Y$ and $X \to X'$ can be completed to a morphism

$$
\begin{array}{ccccccc}
X & \longrightarrow & Y & \longrightarrow & Z & \longrightarrow & \Sigma X \\
\downarrow & & \downarrow & & \| & & \downarrow \\
X' & \longrightarrow & Y' & \longrightarrow & Z & \longrightarrow & \Sigma X'
\end{array}
$$

between exact triangles such that the left-hand square is homotopy cartesian and the composite $Y' \to Z \to \Sigma X$ is a differential.

Remark 1.38. Part of axiom (T1) is that each morphism $X \to Y$ can be completed to an exact triangle $X \to Y \to Z \to \Sigma X$; it follows from the other axioms that the object Z is unique up to isomorphism, though not up to a unique isomorphism.

As a final remark about the axioms of a triangulated category, we note that they are self-dual, so that the opposite category is also triangulated.

1.3.2 Categories of complexes

Let A be an additive category. A *complex* in A is a sequence of morphisms

$$
\cdots \to X^{n-1} \xrightarrow{d^{n-1}} X^n \xrightarrow{d^n} X^{n+1} \to \cdots
$$

such that $d^n \circ d^{n-1} = 0$ for all $n \in \mathbb{Z}$. A morphism $\phi \colon X \to Y$ between complexes consists of morphisms $\phi^n \colon X^n \to Y^n$ with $d_Y^n \circ \phi^n = \phi^{n+1} \circ d_X^n$ for all $n \in \mathbb{Z}$. The complexes form a category which we denote by $\mathsf{C}(\mathsf{A})$.

A morphism $\phi \colon X \to Y$ is *null-homotopic* if there are morphisms $\rho^n \colon X^n \to Y^{n-1}$ such that $\phi^n = d_Y^{n-1} \circ \rho^n + \rho^{n+1} \circ d_X^n$ for all $n \in \mathbb{Z}$. Morphisms $\phi, \psi \colon X \to Y$ are *homotopic* if $\phi - \psi$ is null-homotopic.

The null-homotopic morphisms form an *ideal* $\mathcal{I}$ in $\mathsf{C}(\mathsf{A})$, that is, for each pair X, Y of complexes a subgroup

$$
\mathcal{I}(X,Y) \subseteq \mathrm{Hom}_{\mathsf{C}(\mathsf{A})}(X,Y)
$$

such that any composite $\psi \circ \phi$ of morphisms in $\mathsf{C}(\mathsf{A})$ belongs to $\mathcal{I}$ if ϕ or ψ belongs to $\mathcal{I}$. The *homotopy category* $\mathsf{K}(\mathsf{A})$ is the quotient of $\mathsf{C}(\mathsf{A})$ with respect to this ideal. Thus

$$
\mathrm{Hom}_{\mathsf{K}(\mathsf{A})}(X,Y) = \mathrm{Hom}_{\mathsf{C}(\mathsf{A})}(X,Y)/\mathcal{I}(X,Y)
$$

for every pair of complexes X, Y.

Given any complex X, its *suspension* or *shift* is the complex ΣX with

$$
(\Sigma X)^n = X^{n+1} \quad \text{and} \quad d_{\Sigma X}^n = -d_X^{n+1}.
$$

This yields an equivalence $\Sigma \colon \mathsf{K}(\mathsf{A}) \xrightarrow{\sim} \mathsf{K}(\mathsf{A})$. The *mapping cone* of a morphism $\alpha \colon X \to Y$ of complexes is the complex Z with $Z^n = X^{n+1} \amalg Y^n$ and differential

$$
\begin{bmatrix} -d_X^{n+1} & 0 \\ \alpha^{n+1} & d_Y^n \end{bmatrix}.
$$

The mapping cone fits into a *mapping cone sequence*

$$X \xrightarrow{\alpha} Y \xrightarrow{\beta} Z \xrightarrow{\gamma} \Sigma X$$

which is defined in degree n by the following sequence.

$$X^n \xrightarrow{\alpha^n} Y^n \xrightarrow{\left[\begin{smallmatrix} 0 \\ \mathrm{id} \end{smallmatrix}\right]} X^{n+1} \amalg Y^n \xrightarrow{[-\,\mathrm{id}\ 0]} X^{n+1}.$$

By definition, a triangle in $\mathsf{K}(\mathsf{A})$ is *exact* if it is isomorphic to a mapping cone sequence as above. It is straightforward to verify the axioms (T1)–(T4). Thus $\mathsf{K}(\mathsf{A})$ is a triangulated category.

1.3.3   Exact categories

Let A be an exact category in the sense of Quillen [50]. Thus A is an additive category, together with a distinguished class of sequences

$$0 \to X \xrightarrow{\alpha} Y \xrightarrow{\beta} Z \to 0$$

which are called *exact*. The exact sequences satisfy a number of axioms. In particular, the morphisms α and β in each exact sequence as above form a *kernel-cokernel pair*, that is α is a kernel of β and β is a cokernel of α. A morphism in A which arises as the kernel in some exact sequence is called *admissible mono*; a morphism arising as a cokernel is called *admissible epi*. A full subcategory B of A is *extension-closed* if every exact sequence in A with endterms in B belongs to B.

Remark 1.39. (1) Any abelian category is exact with respect to the class of all short exact sequences.

(2) Any full and extension-closed subcategory B of an exact category A is exact with respect to the class of sequences which are exact in A.

(3) Any small exact category arises, up to an exact equivalence, as a full and extension-closed subcategory of a module category.

1.3.4   Frobenius categories

Let A be an exact category. An object P is called *projective* if the induced map $\mathrm{Hom}_{\mathsf{A}}(P, Y) \to \mathrm{Hom}_{\mathsf{A}}(P, Z)$ is surjective for every admissible epi $Y \to Z$. Dually, an object Q is *injective* if the induced map $\mathrm{Hom}_{\mathsf{A}}(Y, Q) \to \mathrm{Hom}_{\mathsf{A}}(X, Q)$ is surjective for every admissible mono $X \to Y$. The category A has *enough projectives* if every object Z admits an admissible epi $Y \to Z$ with Y projective. And A has *enough injectives* if every object X admits an admissible mono $X \to Y$ with Y injective. Finally, A is called a *Frobenius category*, if A has enough projectives and enough injectives and if both coincide.

Example 1.40. (1) Let A be an additive category. Then A is an exact category with respect to the class of all split exact sequences in A. All objects are projective and injective, and A is a Frobenius category.

(2) Let A be an additive category. The category $C(A)$ of complexes is exact with respect to the class of all sequences $0 \to X \to Y \to Z \to 0$ such that $0 \to X^n \to Y^n \to Z^n \to 0$ is split exact for all $n \in \mathbb{Z}$. A typical projective and injective object is a complex of the form

$$I_A: \quad \cdots \to 0 \to A \xrightarrow{\text{id}} A \to 0 \to \cdots$$

for some A in A. There is an obvious admissible mono $X \to \prod_{n \in \mathbb{Z}} \Sigma^{-n} I_{X^n}$ and also an admissible epi $\coprod_{n \in \mathbb{Z}} \Sigma^{-n-1} I_{X^n} \to X$. Also,

$$\coprod_{n \in \mathbb{Z}} \Sigma^{-n} I_{X^n} \cong \prod_{n \in \mathbb{Z}} \Sigma^{-n} I_{X^n}.$$

Thus $C(A)$ is a Frobenius category. For a conceptual explanation, see Exercise 22 at the end of this chapter.

(3) Let k be a field and A a finite-dimensional self-injective k-algebra. Then injective and projective A-modules coincide. Thus the category of A-modules is a Frobenius category.

1.3.5 The stable category of a Frobenius category

Let A be a Frobenius category. The *stable category* $S(A)$ is by definition the quotient of A with respect to the ideal $\mathcal{I}$ of morphisms which factor through an injective object. Thus the objects of $S(A)$ are the same as in A and

$$\text{Hom}_{S(A)}(X, Y) = \text{Hom}_A(X, Y) / \mathcal{I}(X, Y)$$

for all X, Y in A.

We define a triangulated structure for $S(A)$ as follows. Choose for each X in A an exact sequence

$$0 \to X \to E \to \Sigma X \to 0$$

such that E is injective. If $0 \to X \to E' \to Y \to 0$ is another exact sequence with E' injective, then there is an isomorphism $E \oplus Y \cong E' \oplus \Sigma X$; prove this! Thus, the assignment $X \mapsto \Sigma X$ is well-defined in $S(A)$ and it yields an equivalence $\Sigma \colon S(A) \xrightarrow{\sim} S(A)$. This equivalence serves as suspension. Every exact sequence $0 \to X \to Y \to Z \to 0$ fits into a commutative diagram with exact rows

$$
\begin{array}{ccccccccc}
0 & \longrightarrow & X & \xrightarrow{\alpha} & Y & \xrightarrow{\beta} & Z & \longrightarrow & 0 \\
& & \| & & \downarrow & & \downarrow{\scriptstyle \gamma} & & \\
0 & \longrightarrow & X & \longrightarrow & E & \longrightarrow & \Sigma X & \longrightarrow & 0
\end{array}
$$

such that E is injective. A triangle in $\mathsf{S}(\mathsf{A})$ is by definition *exact* if it isomorphic
to a sequence of morphisms

$$X \xrightarrow{\alpha} Y \xrightarrow{\beta} Z \xrightarrow{\gamma} \Sigma X$$

as above.

Proposition 1.41. *The stable category of a Frobenius category is triangulated.*

Proof. It is easy to verify the axioms, once one observes that every morphism
in $\mathsf{S}(\mathsf{A})$ can be represented by an admissible mono in A. Note that a homotopy
cartesian square can be represented by a pull-back and push-out square. This gives
a proof for (T4′). We refer to [33, §I.2] for details.   □

Example 1.42. (1) Let A be a finite-dimensional and self-injective algebra. Then
the category of A-modules $\mathsf{Mod}\,A$ is a Frobenius category, and we write $\mathsf{StMod}\,A$
for the stable category $\mathsf{S}(\mathsf{Mod}\,A)$.

(2) The category of complexes $\mathsf{C}(\mathsf{A})$ of an additive category A is a Frobenius
category with respect to the degreewise split exact sequences. The morphisms
factoring through an injective object are precisely the null-homotopic morphisms.
Thus the stable category of $\mathsf{C}(\mathsf{A})$ coincides with the homotopy category $\mathsf{K}(\mathsf{A})$. Note
that the triangulated structures which have been defined via mapping cones and
via exact sequences in $\mathsf{C}(\mathsf{A})$ coincide.

1.3.6   Exact and cohomological functors

An *exact functor* $\mathsf{T} \to \mathsf{U}$ between triangulated categories is a pair (F, η) consisting
of an additive functor $F \colon \mathsf{T} \to \mathsf{U}$ and a natural isomorphism $\eta \colon F \circ \Sigma_{\mathsf{T}} \to \Sigma_{\mathsf{U}} \circ F$
such that for every exact triangle $X \xrightarrow{\alpha} Y \xrightarrow{\beta} Z \xrightarrow{\gamma} \Sigma X$ in T the triangle

$$FX \xrightarrow{F\alpha} FY \xrightarrow{F\beta} FZ \xrightarrow{\eta_X \circ F\gamma} \Sigma(FX)$$

is exact in U.

Example 1.43. An additive functor $\mathsf{A} \to \mathsf{B}$ induces an exact functor $\mathsf{K}(\mathsf{A}) \to \mathsf{K}(\mathsf{B})$.

A functor $\mathsf{T} \to \mathsf{A}$ from a triangulated category T to an abelian category A is
cohomological if it sends each exact triangle in T to an exact sequence in A.

Example 1.44. For each object X in T, the representable functors

$$\mathrm{Hom}_{\mathsf{T}}(X, -) \colon \mathsf{T} \to \mathsf{Ab} \quad \text{and} \quad \mathrm{Hom}_{\mathsf{T}}(-, X) \colon \mathsf{T}^{\mathrm{op}} \to \mathsf{Ab}$$

into the category Ab of abelian groups are cohomological functors.

1.3.7 Thick subcategories

Let T be a triangulated category. A non-empty full subcategory S is a *triangulated subcategory* if the following conditions hold.

(S1) $\Sigma^n X \in S$ for all $X \in S$ and $n \in \mathbb{Z}$.

(S2) Let $X \to Y \to Z \to \Sigma X$ be an exact triangle in T. If two objects from $\{X, Y, Z\}$ belong to S, then so does the third.

A triangulated subcategory S is *thick* if in addition the following condition holds.

(S3) Every direct factor of an object in S belongs to S, that is, a decomposition $X = X' \amalg X''$ for $X \in S$ implies $X' \in S$.

A triangulated subcategory S inherits a canonical triangulated structure from T.

Example 1.45. Let $F\colon T \to U$ be an exact functor between triangulated categories. Then the full subcategory $\operatorname{Ker} F$, called the *kernel* of F, consisting of the objects annihilated by F forms a thick subcategory of T.

1.3.8 Derived categories

Let A be an abelian category. Given a complex

$$\cdots \to X^{n-1} \xrightarrow{d^{n-1}} X^n \xrightarrow{d^n} X^{n+1} \to \cdots$$

in A, the *cohomology* in degree n is by definition the object

$$H^n X = \operatorname{Ker} d^n / \operatorname{Im} d^{n-1}.$$

A morphism $\phi\colon X \to Y$ of complexes induces, for each $n \in \mathbb{Z}$, a morphism

$$H^n\phi\colon H^n X \to H^n Y \, ;$$

if these are all isomorphisms, then ϕ is said to be a *quasi-isomorphism*. Note that morphisms $\phi, \psi\colon X \to Y$ are homotopic, then $H^n\phi = H^n\psi$ for all n.

The *derived category* D(A) of A is obtained from K(A) by formally inverting all quasi-isomorphisms. To be precise, one defines

$$D(A) = K(A)[S^{-1}]$$

as the localisation of K(A) with respect to the class S of all quasi-isomorphisms.

Proposition 1.46. *The derived category* D(A) *carries a unique triangulated structure such that the canonical functor* K(A) $\to$ D(A) *is exact.* $\qquad\square$

We identify any object X in A with the complex having X concentrated in degree zero. This yields a functor $\mathsf{A} \to \mathsf{D}(\mathsf{A})$ which induces for all objects X, Y in A and $n \in \mathbb{Z}$ an isomorphism

$$\mathrm{Ext}^n_{\mathsf{A}}(X, Y) \xrightarrow{\sim} \mathrm{Hom}_{\mathsf{D}(\mathsf{A})}(X, \Sigma^n Y).$$

For example, the functor sends each exact sequence $\eta\colon 0 \to A \xrightarrow{\alpha} B \xrightarrow{\beta} C \to 0$ in A to an exact triangle $A \xrightarrow{\alpha} B \xrightarrow{\beta} C \xrightarrow{\gamma} \Sigma A$. The above isomorphism maps the class in $\mathrm{Ext}^1_{\mathsf{A}}(C, A)$ representing η to $\gamma\colon C \to \Sigma A$.

1.3.9   Compact objects

Let T be a triangulated category and suppose that T admits set-indexed coproducts. A *localising subcategory* of T is a full triangulated subcategory that is closed under taking set-indexed coproducts. A localising subcategory is also a thick subcategory; see, for example, [36, Lemma 1.4.9]. We write $\mathrm{Loc}_{\mathsf{T}}(\mathsf{C})$ for the smallest localising subcategory containing a given class of objects C in T, and call it the localising subcategory *generated* by C. In the same vein, we write $\mathrm{Thick}_{\mathsf{T}}(\mathsf{C})$ for the smallest thick subcategory containing C, and call it the thick subcategory generated by C.

An object X in T is *compact* if the functor $\mathrm{Hom}_{\mathsf{T}}(X, -)$ commutes with all coproducts. This means that each morphism $X \to \coprod_{i \in I} Y_i$ in T factors through $X \to \coprod_{i \in J} Y_i$ for some finite subset $J \subseteq I$. We write T^c for the full subcategory of compact objects in T. Note that T^c is a thick subcategory of T.

The category T is *compactly generated* if it is generated by a set of compact objects, that is, $\mathsf{T} = \mathrm{Loc}_{\mathsf{T}}(\mathsf{C})$ for some set $\mathsf{C} \subseteq \mathsf{T}^c$. The following result provides a useful criterion for compact generation. The proof uses the Brown representability theorem, which we will learn about in Section 2.2.

Proposition 1.47. *Let C be a set of compact objects of T. Then $\mathrm{Loc}_{\mathsf{T}}(\mathsf{C}) = \mathsf{T}$ if and only if for each non-zero object $X \in \mathsf{T}$ there are $C \in \mathsf{C}$ and $n \in \mathbb{Z}$ such that $\mathrm{Hom}_{\mathsf{T}}(\Sigma^n C, X) \neq 0$.*

Proof. Assume $\mathrm{Loc}_{\mathsf{T}}(\mathsf{C}) = \mathsf{T}$ and fix an object $X \in \mathsf{T}$. The objects $V \in \mathsf{T}$ satisfying $\mathrm{Hom}_{\mathsf{T}}(\Sigma^n V, X) = 0$ for all $n \in \mathbb{Z}$ form a localising subcategory of T. If this localising subcategory contains C, then $X = 0$.

As to the converse, Corollary 2.13 yields that the inclusion $\mathrm{Loc}_{\mathsf{T}}(\mathsf{C}) \to \mathsf{T}$ admits a right adjoint; we denote this by Γ. Given any object $X \in \mathsf{T}$, this yields a universal morphism $\Gamma X \to X$. Completing this to an exact triangle $\Gamma X \to X \to X' \to$ produces an object X' satisfying $\mathrm{Hom}_{\mathsf{T}}(V, X') = 0$ for all $V \in \mathrm{Loc}_{\mathsf{T}}(\mathsf{C})$. Thus $X' = 0$ and therefore X belongs to $\mathrm{Loc}_{\mathsf{T}}(\mathsf{C})$. $\square$

Example 1.48. (1) Let A be any ring and $\mathsf{D}(\mathsf{Mod}\, A)$ its derived category. Since $\mathrm{Hom}_{\mathsf{D}(\mathsf{Mod}\, A)}(\Sigma^n A, X) = H^{-n}(X)$, it follows that A viewed as complex concentrated in degree zero is a compact object; it is also a generator, by Proposition 1.47.

Thus the derived category is compactly generated. The compact objects are described in Theorem 2.2.

(2) Let A be a finite-dimensional and self-injective algebra. Then $\mathsf{Mod}\,A$ is a Frobenius category and the corresponding stable category $\mathsf{StMod}\,A$ is compactly generated. An object is compact if and only if it is isomorphic to a finitely generated A-module. Thus the inclusion $\mathsf{mod}\,A \to \mathsf{Mod}\,A$ induces an equivalence $\mathsf{stmod}\,A \xrightarrow{\sim} (\mathsf{StMod}\,A)^c$.

Indeed, $\mathsf{Mod}\,A$ with exact structure given by exact sequences of modules is a Frobenius category, with projectives the projective A-modules; see Example 1.42. Its stable category is thus triangulated with suspension Ω^{-1}, by the discussion in Section 1.3.5. The simple A-modules form a set of compact generators. This follows from Proposition 1.47. The thick subcategory generated by all simple A-modules coincides with $\mathsf{stmod}\,A$, and this yields the description of the compact objects of $\mathsf{StMod}\,A$, by Theorem 1.49 below.

The result below, attributed to Ravenel [51], is extremely useful in identifying compact objects in triangulated categories; see [45, Lemma 2.2] for a proof.

Theorem 1.49. *Let C and D be compact objects in a triangulated category T. If D is in $\mathrm{Loc}_{\mathsf{T}}(C)$, then it is already in $\mathrm{Thick}_{\mathsf{T}}(C)$. In particular, if C is a compact generator for T, then the class of compact objects in T is precisely $\mathrm{Thick}_{\mathsf{T}}(C)$.* $\square$

1.4 Exercises

In the following exercises k is a field, G a finite group, and M, N are kG-modules. Keep in mind that the kG action on $M \otimes_k N$ is via the diagonal. In what follows char k denotes the characteristic of k.

(1) Let H be a finite group. Prove that the canonical homomorphisms $G \to G \times H$ and $H \to G \times H$ of groups induce an isomorphism of k-algebras:

$$kG \otimes_k kH \xrightarrow{\cong} k[G \times H].$$

(2) Let $G = \langle g_1, \ldots, g_r \rangle \cong (\mathbb{Z}/p)^r$ and set $x_i = g_i - 1$, in kG. Prove that if char $k = p$, then kG is isomorphic as a k-algebra to

$$k[x_1, \ldots, x_r]/(x_1^p, \ldots, x_r^p).$$

(3) Describe the centre of the ring kG, for a general group G.

(4) Let $\pi \colon kG \to k$ be the k-linear map defined on the basis G by $\pi(1) = 1$ and $\pi(g) = 0$ for $g \neq 1$. Verify that the following map is a kG-linear isomorphism.

$$kG \to \mathrm{Hom}_k(kG, k) \quad \text{where } g \mapsto [h \mapsto \pi(g^{-1}h)].$$

This proves that kG is a self-injective algebra.

(5) Verify that the following maps are kG-linear isomorphisms.

$$M \to k \otimes_k M \qquad \text{where } m \mapsto 1 \otimes m\,;$$
$$M{\downarrow_1}{\uparrow}^G \to kG \otimes_k M \quad \text{where } g \otimes m \mapsto g \otimes gm\,.$$

The second isomorphism implies that $kG \otimes_k M$ is a free kG-module.

(6) Verify that the following maps are kG-linear monomorphisms:

$$M \to kG \otimes_k M \quad \text{where } m \mapsto \sum_{g \in G} g \otimes m\,;$$
$$M \to M{\downarrow_1}{\uparrow}^G \qquad \text{where } m \mapsto \sum_{g \in G} g \otimes g^{-1}m\,.$$

Since $M{\downarrow_1}{\uparrow}^G$ is free, it follows that each module embeds into a free one, in a canonical way.

(7) Prove that a kG-module M is projective if and only if it is injective. Hint: use (4)–(6). It is also true that M is projective if and only if it is flat.

(8) Let H be a subgroup of G. Prove that kG is free as a kH-module, both on the left and on the right, and describe bases.

(9) Let G be a finite p-group and char $k = p$. For any non-zero element $m \in M$ the $\mathbb{F}_p$-subspace of M spanned by $\{gm \mid g \in G\}$ is finite-dimensional, and so has p^n elements for some n. Show that some non-zero element of this set is fixed by G. Deduce that the trivial module is the only simple kG-module.

(10) Let $G = (\mathbb{Z}/p)^r$ and char $k = p$. Describe $J(kG)$, the Jacobson radical of kG, and show that $J(kG)/J^2(kG)$ is a vector space of dimension r over k. Prove that there is a natural isomorphism of k-vector spaces

$$H^1(G, k) \cong \mathrm{Hom}_k(J(kG)/J^2(kG), k)\,.$$

(11) Let $G = \langle g \mid g^{p^n} = 1 \rangle \cong \mathbb{Z}/p^n$ with $n > 1$ and char $k = p$. Use Jordan canonical form to show that a finitely generated kG-module is free if and only if its restriction to the subgroup

$$H = \langle g^{p^{n-1}} \rangle \cong \mathbb{Z}/p$$

is free. This is a case of Chouinard's theorem.

In the following exercises, assume the field k **is algebraically closed**.

(12) Write $(\mathbb{Z}/2)^2 = \langle g_1, g_2 \rangle$ and assume char $k = 2$. For each $\lambda \in k$, compute the rank variety of the following module M_λ:

$$g_1 \mapsto \begin{pmatrix} 1 & 0 \\ 1 & 1 \end{pmatrix}, \qquad g_2 \mapsto \begin{pmatrix} 1 & 0 \\ \lambda & 1 \end{pmatrix}.$$

Deduce that the M_λ are non-isomorphic for different values of λ.

(13) Write $(\mathbb{Z}/3)^2 = \langle g_1, g_2 \rangle$ and assume char $k = 3$. For each $\lambda \in k$, compute the rank variety of the following module M_λ:

$$g_1 \mapsto \begin{pmatrix} 1 & 0 & 0 \\ 1 & 1 & 0 \\ 0 & 1 & 1 \end{pmatrix}, \qquad g_2 \mapsto \begin{pmatrix} 1 & 0 & 0 \\ \lambda & 1 & 0 \\ 0 & \lambda & 1 \end{pmatrix}.$$

Prove that the M_λ are non-isomorphic for different values of λ.

(14) Generalise the last question to $(\mathbb{Z}/p)^2$ in characteristic p, and deduce that there are infinitely many isomorphism classes of p-dimensional modules.

(15) Write $(\mathbb{Z}/2)^4 = \langle g_1, g_2, g_3, g_4 \rangle$ and assume char $k = 2$. Compute the rank variety of the following module:

$$g_1 \mapsto \begin{pmatrix} 1 & 0 & 0 & 0 \\ 0 & 1 & 0 & 0 \\ 1 & 0 & 1 & 0 \\ 0 & 0 & 0 & 1 \end{pmatrix}, \qquad g_2 \mapsto \begin{pmatrix} 1 & 0 & 0 & 0 \\ 0 & 1 & 0 & 0 \\ 0 & 1 & 1 & 0 \\ 0 & 0 & 0 & 1 \end{pmatrix},$$

$$g_3 \mapsto \begin{pmatrix} 1 & 0 & 0 & 0 \\ 0 & 1 & 0 & 0 \\ 0 & 0 & 1 & 0 \\ 1 & 0 & 0 & 1 \end{pmatrix}, \qquad g_4 \mapsto \begin{pmatrix} 1 & 0 & 0 & 0 \\ 0 & 1 & 0 & 0 \\ 0 & 0 & 1 & 0 \\ 0 & 1 & 0 & 1 \end{pmatrix}.$$

(16) Let $G = \langle g_1, \ldots, g_r \rangle \cong (\mathbb{Z}/2)^r$ for some $r > 1$ and assume char $k = 2$. Set $K = k(t_1, \ldots, t_r)$, a transcendental extension of k and let $M = K \oplus K$ as a k-vector space, with G-action given by

$$g_i \mapsto \begin{pmatrix} I & 0 \\ t_i I & I \end{pmatrix}.$$

Prove that M is not free, but that its restriction to every cyclic shifted subgroup of G is free.

Hint for the first part: if $i \neq j$, then $(g_i - 1)(g_j - 1)$ acts as zero.

(17) Let T be a triangulated category. Prove that for each object X in T, the contravariant representable functor $\mathrm{Hom}_{\mathsf{T}}(-, X)$ and the covariant representable functor $\mathrm{Hom}_{\mathsf{T}}(X, -)$ are cohomological.

(18) Prove that the class of compact objects in a triangulated category admitting arbitrary coproducts forms a thick subcategory.

(19) Prove that in a triangulated category a coproduct of exact triangles is exact.

(20) Prove that given an adjoint pair of functors between triangulated categories, one of the functors is exact if and only if the other is.

(21) Let A be an additive category. Prove that A is a Frobenius category with exact structure given by the split short exact sequences in A.

(22) Let A be an additive category and C(A) the category of complexes over A, with morphisms the degree zero chain maps. Prove that C(A) is a Frobenius category, with exact structure given by the degreewise split exact sequences.

Hint: One can deduce this from general principles as follows. The functor $F \colon \mathsf{C}(\mathsf{A}) \to \prod_{n \in \mathbb{Z}} \mathsf{A}$ forgetting the differential is exact. Here we use the exact structure on $\prod_{n \in \mathbb{Z}} \mathsf{A}$ coming from the additive structure. The functor F has a left adjoint F_λ and a right adjoint F_ρ. Thus F_λ sends projectives to projectives, while F_ρ sends injectives to injectives. For each complex X, the counit $F_\lambda F X \to X$ is an admissable epi, and the unit $X \to F_\rho F X$ is an admissable mono.

(23) Let $F \colon \mathsf{T} \to \mathsf{U}$ be an exact functor between triangulated categories that admit set-indexed coproducts. Suppose also that T is compactly generated by a compact object C. Then F is an equivalence of triangulated categories if and only if F induces a bijection

$$\operatorname{Hom}_{\mathsf{T}}(C, \Sigma^n C) \xrightarrow{\sim} \operatorname{Hom}_{\mathsf{U}}(FC, \Sigma^n FC)$$

for all $n \in \mathbb{Z}$, and $\operatorname{Loc}_{\mathsf{U}}(FC) = \mathsf{U}$.

Hint: See [11, Lemma 4.5].

2 Tuesday

The highlights of this chapter are Hopkins' theorem on perfect complexes over commutative noetherian rings, which is the content of Section 2.1, and its analogue in modular representation theory, proved by Benson, Carlson, and Rickard; this appears in Section 2.3. This requires a discussion of appropriate notions of support; for commutative rings, this is based on the material from Appendix A, while for modules over group algebras, one requires more sophisticated tools, from homotopy theory, and these are discussed in Section 2.2.

2.1 Perfect complexes over commutative rings

In this lecture A will denote a commutative noetherian ring. Good examples to bear in mind are polynomial rings over fields, their quotients, and localisations. We write $\mathsf{Mod}\,A$ for the category of A-modules, and $\mathsf{mod}\,A$ for its full subcategory of finitely generated A-modules.

Keeping with the convention in these notes, complexes of A-modules will be upper graded:

$$\cdots \to M^{i-1} \xrightarrow{d^{i-1}} M^i \xrightarrow{d^i} M^{i+1} \to \cdots$$

The nth *suspension* of M is denoted $\Sigma^n M$; thus for each $i \in \mathbb{Z}$ one has

$$(\Sigma^n M)^i = M^{n+i} \quad \text{and} \quad d^{\Sigma^n M} = (-1)^n d^M .$$

We write $H^*(M)$ for the total cohomology $\bigoplus_{n \in \mathbb{Z}} H^n(M)$ of such a complex. It is viewed as a graded module over the ring A.

Let $\mathsf{D}(A)$ denote the derived category of the category of A-modules, viewed as a triangulated category with Σ as the translation functor. Let $\mathsf{D}^b(\mathsf{mod}\,A)$ denote the bounded derived category of finitely generated A-modules. Sometimes it is convenient to identify $\mathsf{D}^b(\mathsf{mod}\,A)$ with the full subcategory

$$\{M \in \mathsf{D}(A) \mid \text{the } A\text{-module } H^*(M) \text{ is finitely generated}\}$$

of $\mathsf{D}(A)$ consisting of complexes with finitely generated cohomology.

2.1.1   Perfect complexes

We focus on thick subcategories of $\mathsf{D}^b(\mathrm{mod}\,A)$. Recall that a non-empty subcat-egory $\mathsf{C} \subseteq \mathsf{D}^b(\mathrm{mod}\,A)$ is said to be *thick* if it is closed under mapping cones, suspensions, and retracts.

Our interest in thick subcategories of $\mathsf{D}^b(\mathrm{mod}\,A)$ stems from the fact that they are kernels of exact and cohomological functors. In addition most "reason-able" homological properties of complexes give rise to thick subcategories; the reader may take this as the definition of a reasonable property. A central problem which motivates the techniques and results discussed in these lectures is:

Problem: Classify the thick subcategories of $\mathsf{D}^b(\mathrm{mod}\,A)$.

In the course of these lectures the reader will encounter solutions to this problem for important classes of rings. In this lecture, we present a classification theorem for a small piece of the bounded derived category of a commutative ring.

Definition 2.1. Let M be a complex of A-modules. We write $\mathrm{Thick}(M)$ for the smallest (with respect to inclusion) thick subcategory of $\mathsf{D}(A)$ containing M.

An intersection of thick subcategories is again thick, so $\mathrm{Thick}(M)$ equals the intersection of all thick subcategories of $\mathsf{D}(A)$ containing M. The objects in $\mathrm{Thick}(M)$ are the complexes that can be *finitely built* out of M using finite direct sums, suspensions, exact triangles, and retracts. See [2, §2] for a more precise and constructive definition.

When $H^*(M)$ is finitely generated, $\mathrm{Thick}(M)$ is contained in $\mathsf{D}^b(\mathrm{mod}\,A)$, for the latter is a thick subcategory of $\mathsf{D}(A)$.

Theorem 2.2. *Let M be in $\mathsf{D}(A)$. The following conditions are equivalent:*

(1) *M is in $\mathrm{Thick}(A)$.*

(2) *M is isomorphic in $\mathsf{D}(A)$ to a bounded complex $0 \to P^i \to \cdots \to P^s \to 0$ of finitely generated projective A-modules.*

(3) *M is a compact object in $\mathsf{D}(A)$.*

In commutative algebra any bounded complex of projective modules, as in (2), is said to be *perfect*. We apply this terminology to any complex isomorphic to such a complex, and so, given the result above, speak of $\mathrm{Thick}(A)$ as the sub-category of perfect complexes in $\mathsf{D}(A)$.

Proof. Note that $\mathrm{Hom}_{\mathsf{D}(A)}(A, X) = H^0(X)$ for any complex X of A-modules.

(1) $\implies$ (3) It follows from the identification above that A is a compact ob-ject, for $H^0(-)$ commutes with arbitrary coproducts. The class of compact objects in any triangulated category, in particular, in $\mathsf{D}(A)$, form a thick subcategory – this was an exercise in Chapter 1. Thus $\mathrm{Thick}(A)$ consists of compact objects.

(3) $\implies$ (1) Since A is compact in $\mathsf{D}(A)$ and generates it (see Example 1.48), this implication follows from Theorem 1.49.

(1) $\implies$ (2) Using the Horseshoe Lemma [55, Lemma 2.2.8], it is a straightforward exercise to prove that the complexes as in (2) form a thick subcategory. Since A is evidently contained in it, so is all of Thick(A).

(2) $\implies$ (1) Since Thick(A) is closed under finite sums, it contains A^n for each $n \geq 1$. Since it is closed also under direct summands, it contains all finitely generated projectives, and also their shifts.

Any complex of the form $P := 0 \to P^i \to \cdots \to P^s \to 0$ fits into an exact sequence of A-modules

$$0 \to \Sigma^{-s} P^s \to P \to P^{\leqslant s-1} \to 0$$

Thus, an induction on the number of non-zero components of P yields that P, and hence anything isomorphic to it, is in Thick(A). $\qquad\square$

Remark 2.3. There is an inclusion Thick(A) $\subseteq$ $\mathsf{D}^b(\text{mod } A)$. In view of the implication (1) $\implies$ (2) of the preceding result, equality holds if and only if each M in $\mathsf{D}^b(\text{mod } A)$ has a finite projective resolution. The latter condition is equivalent to the statement that the ring A is *regular*, by a theorem of Auslander, Buchsbaum, and Serre; see [43, Section 19] for the definition of a regular ring, and for a proof of that theorem. We note only that fields are regular rings, and the regularity property is inherited by localisations and polynomial extensions.

We are heading towards a description of the thick subcategories of Thick(A). This involves a notion of support for complexes; what is described below is an extension of "big support" of modules defined in Appendix A.

2.1.2 Support

Recall that the *Zariski spectrum* of the ring A is the set

$$\text{Spec } A = \{\mathfrak{p} \subseteq A \mid \mathfrak{p} \text{ is a prime ideal}\}$$

with the Zariski topology. The closed sets are $\mathcal{V}(\mathfrak{a}) = \{\mathfrak{p} \in \text{Spec } A \mid \mathfrak{p} \supseteq \mathfrak{a}\}$, where $\mathfrak{a}$ is an ideal in A. For each M in $\mathsf{D}(A)$ we set

$$\text{Supp}_A M = \{\mathfrak{p} \in \text{Spec } A \mid H^*(M_{\mathfrak{p}}) \neq 0\}.$$

Note that passing to cohomology commutes with localisation: $H^*(M_{\mathfrak{p}}) = H^*(M)_{\mathfrak{p}}$. The following properties are readily verified.

(1) For each $M \subset \mathsf{D}^b(\text{mod } A)$ one has

$$\text{Supp}_A M = \text{Supp}_A H^*(M) = \mathcal{V}(\text{ann}_A H^*(M)).$$

In particular, $\text{Supp}_A M$ is a closed subset of Spec A.

(2) $\text{Supp}_A M = \varnothing$ if and only if $H^*(M) = 0$, that is to say, if $M = 0$ in $\mathsf{D}(A)$.

(3) $\text{Supp}_A(M \oplus N) = \text{Supp}_A M \bigcup \text{Supp}_A N$.

(4) Given an exact triangle $L \to M \to N \to \Sigma L$ in $\mathsf{D}(\mathrm{mod}\, A)$ one has

$$\mathrm{Supp}_A M \subseteq \mathrm{Supp}_A L \cup \mathrm{Supp}_A N .$$

Moreover, $\mathrm{Supp}_A(\Sigma M) = \mathrm{Supp}_A M$.

(5) $\mathrm{Supp}_A(M \otimes^{\mathbf{L}}_A N) = \mathrm{Supp}_A M \cap \mathrm{Supp}_A N$ for all M, N in $\mathsf{D}^{\mathsf{b}}(\mathrm{mod}\, A)$.

Compare with properties of support for group representations in Section 1.1.7.
    A routine verification, using the properties of support above, yields:

Lemma 2.4. *For any subset $\mathcal{U}$ of* $\mathrm{Spec}\, A$, *the full subcategory*

$$\mathsf{C}_{\mathcal{U}} = \{M \in \mathsf{D}^{\mathsf{b}}(\mathrm{mod}\, A) \mid \mathrm{Supp}_A M \subseteq \mathcal{U}\}$$

of $\mathsf{D}^{\mathsf{b}}(\mathrm{mod}\, A)$ *is a thick subcategory.* □

2.1.3   Koszul complexes

Let a be an element in the ring A. The *Koszul complex* on a is the complex

$$\cdots \to 0 \to A \xrightarrow{a} A \to 0 \cdots \to$$

of A-modules, where the non-zero components are in degree -1 and 0; remember
that the grading is cohomological. One can view it as the *mapping cone* of the
morphism of complexes $A \xrightarrow{a} A$ where A is viewed as a complex, in the usual way.
    The Koszul complex on a with coefficients in a complex M is the mapping
cone of the morphism $M \xrightarrow{a} M$, namely, the graded module

$$\Sigma M \oplus M \quad \text{with differential} \quad \begin{bmatrix} -d^M & 0 \\ a & d^M \end{bmatrix}.$$

In anticipation of analogous constructions in triangulated categories we denote
this complex $M /\!/ a$. One then has a *mapping cone exact sequence* of complexes

$$0 \to M \to M /\!/ a \to \Sigma M \to 0 .$$

The Koszul complex on a sequence of elements $\boldsymbol{a} = a_1, \ldots, a_n$ in A is defined by
iterating this construction: $M /\!/ \boldsymbol{a} = M_n$, where $M_0 = M$ and

$$M_i = M_{i-1} /\!/ a_i \quad \text{for } i \geq 1.$$

It is not hard to verify that there is an isomorphism of complexes

$$M /\!/ \boldsymbol{a} \cong (M /\!/ a_1) \otimes_A \cdots \otimes_A (M /\!/ a_n) .$$

Indeed, this is one way to construct the Koszul complex; see [43, §16] for details.
    Finally, given an ideal $\mathfrak{a}$ of A, we let $M /\!/ \mathfrak{a}$ denote a Koszul complex on some
generating set $a_1, \ldots, a_n$ of $\mathfrak{a}$. It does depend on $\boldsymbol{a}$; see, however, Lemma 3.11.

Proposition 2.5. *Let* $\mathfrak{a}$ *be an ideal in* A *and* M *a complex of* A-*modules. The following statements hold.*

(1) $M/\!\!/\mathfrak{a}$ *is in* $\text{Thick}(M)$*; in particular, if* M *is perfect, so is* $M/\!\!/\mathfrak{a}$*.*

(2) $\text{Supp}_A(M/\!\!/\mathfrak{a}) = \text{Supp}_A M \cap \mathcal{V}(\mathfrak{a})$ *when* M *is in* $\mathsf{D}^{\mathrm{b}}(\text{mod}\, A)$*.*

(3) $\text{Supp}_A(A/\!\!/\mathfrak{a}) = \mathcal{V}(\mathfrak{a})$*.*

Proof. It is clear from the mapping cone exact sequence that M finitely builds $M/\!\!/a$, for any element $a \in A$. An iteration yields (1). Moreover, (3) is a special case of (2).

As to (2), it suffices to consider the case of one element a. Then the mapping cone sequence yields in cohomology an exact sequence of graded R-modules:

$$0 \to H^*(M)/aH^*(M) \to H^*(M/\!\!/a) \to (0 :_{H^*(M)} a) \to 0.$$

The module on the right is the submodule of $H^*(M)$ annihilated by (a); in particular, it is supported on $\text{Supp}_R M \cap \mathcal{V}(a)$. Moreover, since $H^*(M)$ is finitely generated, Nakayama's Lemma yields the first equality below:

$$\text{Supp}_R(H^*(M)/aH^*(M)) = \text{Supp}_R H^*(M) \cap \mathcal{V}(a) = \text{Supp}_R M \cap \mathcal{V}(a).$$

Thus the exact sequence above yields $\text{Supp}_R H^*(M/\!\!/a) = \text{Supp}_R M \cap \mathcal{V}(a)$. $\qquad\square$

2.1.4 The theorem of Hopkins

We are now ready to discuss Hopkins' theorem on perfect complexes.

Lemma 2.6. *Let* N *be a complex in* $\mathsf{D}^{\mathrm{b}}(\text{mod}\, A)$*. One then has* $\text{Supp}_A M \subseteq \text{Supp}_A N$ *for each* M *in* $\text{Thick}(N)$*.*

Proof. Apply Lemma 2.4 with $\mathcal{U} = \text{Supp}_A N$, noting that N is in $\mathsf{C}_{\mathcal{U}}$. $\qquad\square$

The converse of the preceding result does not hold in general.

Example 2.7. Let k be a field and set $A = k[x]/(x^2)$. One then has

$$\text{Supp}_A k = \text{Spec}\, A = \text{Supp}_A A\,;$$

however, k is not in $\text{Thick}(A)$ because, for example, k does not admit a finite projective resolution.

It is a remarkable result, proved by Hopkins [35], see also Neeman [44], that for perfect complexes Lemma 2.6 has a perfect converse.

Theorem 2.8 (Hopkins). *If* M *and* N *are perfect complexes and there is an inclusion* $\text{Supp}_A M \subseteq \text{Supp}_A N$*, then* M *is in* $\text{Thick}(N)$*.* $\qquad\square$

A proof of this result, following an idea of Neeman, will be given in Section 5.1. For other proofs see [35], [44], and also [38].

A caveat: In Hopkins' theorem it is crucial that both M and N are perfect, as one can see from Example 2.7. Support over A cannot detect if one complex is built out of another, even for complexes in $\mathsf{D}^{\mathsf{b}}(\mathrm{mod}\,A)$; see the exercises at the end of this chapter.

Corollary 2.9. *When M is a perfect complex,* $\mathrm{Thick}(M) = \mathrm{Thick}(A/\!\!/\mathfrak{a})$ *for any ideal $\mathfrak{a}$ with $\mathcal{V}(\mathfrak{a}) = \mathrm{Supp}_A M$.*   □

We now explain how Hopkins' theorem gives a classification of the thick subcategories of perfect complexes. Recall that a subset $\mathcal{V}$ of $\mathrm{Spec}\,A$ is said to be *specialisation closed* if it has the property that if $\mathfrak{q} \supseteq \mathfrak{p}$ are prime ideals in A and $\mathfrak{p}$ is in $\mathcal{V}$, then so is $\mathfrak{q}$. This is equivalent to the condition that $\mathcal{V}$ is an arbitrary union of closed sets. For any subcategory C of $\mathsf{D}^{\mathsf{b}}(A)$, we set

$$\mathrm{Supp}_A \mathsf{C} = \bigcup_{M \in \mathsf{C}} \mathrm{Supp}_A M \,.$$

This is a specialisation closed subset of $\mathrm{Spec}\,R$, as each $\mathrm{Supp}_A M$ is closed.

Theorem 2.10. *The assignment $\mathsf{C} \mapsto \mathrm{Supp}_A \mathsf{C}$ gives a bijection*

$$\left\{ \begin{array}{c} \textit{Thick subcategories} \\ \textit{of perfect complexes} \end{array} \right\} \longleftrightarrow \left\{ \begin{array}{c} \textit{Specialization closed} \\ \textit{subsets of } \mathrm{Spec}\,A \end{array} \right\}.$$

Its inverse assigns $\mathcal{V}$ to the subcategory $\mathsf{C}_{\mathcal{V}} = \{M \in \mathrm{Thick}(A) \mid \mathrm{Supp}_A M \subseteq \mathcal{V}\}$.

Proof. Let C be a thick subcategory of $\mathrm{Thick}(A)$ and set $\mathcal{V} = \mathrm{Supp}_A \mathsf{C}$. It is clear that $\mathsf{C} \subseteq \mathsf{C}_{\mathcal{V}}$. For equality, it remains to prove that any perfect complex M satisfying $\mathrm{Supp}_A M \subseteq \mathcal{V}$ is in C. This is a consequence of Theorem 2.8.

Indeed, for M as above, $\mathrm{Supp}_A M$ is a closed set in $\mathrm{Spec}\,A$, hence one can find finitely many complexes in C, say $N_1, \dots, N_k$, such that there is an inclusion

$$\mathrm{Supp}_A M \subseteq \bigcup_{i=1}^{k} \mathrm{Supp}_A(N_i)$$

$$= \mathrm{Supp}_A N \,, \quad \text{where } N = \bigoplus_{i=1}^{k} N_i.$$

The equality holds by properties of support. Since N is also perfect, one obtains $M \in \mathrm{Thick}(N)$, by Theorem 2.8. It remains to note that $\mathrm{Thick}(N) \subseteq \mathsf{C}$, for the latter is a thick subcategory and contains N.

Conversely, for any specialisation closed subset $\mathcal{V}$ of $\mathrm{Spec}\,A$, there is evidently an inclusion $\mathrm{Supp}_A \mathsf{C}_{\mathcal{V}} \subseteq \mathcal{V}$, while equality holds because, for any $\mathfrak{p} \in \mathcal{V}$, the Koszul complex $A/\!\!/\mathfrak{p}$ satisfies:

$$\mathrm{Supp}_A(A/\!\!/\mathfrak{p}) = \mathcal{V}(\mathfrak{p}) \subseteq \mathcal{V} \,,$$

and hence $A/\!\!/\mathfrak{p}$ is in $\mathsf{C}_{\mathcal{V}}$.   □

2.2 Brown representability and localisation

In this section we introduce some technical tools that are important for the rest of this seminar. The principal one is the Brown representability which provides an abstract method for constructing objects in a triangulated category. For this it is important to work in a triangulated category that admits set-indexed coproducts.

Localisation is a method to invert morphisms in a category. If the category is triangulated, then each morphism $\sigma \colon X \to Y$ fits into an exact triangle $X \xrightarrow{\sigma} Y \to Z \to \Sigma X$, and an exact functor sends σ to an invertible morphism if and only if the cone Z is annihilated. Thus we can think of a localisation functor either as a functor that inverts certain morphisms or as one that annihilates certain objects. The objects that are killed by a localisation functor form a localising subcategory.

The first systematic treatment of localisation can be found in [32]. We follow closely the exposition in [10, §3] and refer to [42] for further details.

2.2.1 Brown representability

The following result is known as *Brown representability theorem* and is due to Keller [40] and Neeman [46]; it is a variation of a classical theorem of Brown [19] from homotopy theory.

Theorem 2.11 (Brown). *Let* T *be a compactly generated triangulated category. For a functor* $H \colon \mathsf{T}^{\mathrm{op}} \to \mathsf{Ab}$ *the following are equivalent.*

(1) *The functor* H *is cohomological and preserves set-indexed coproducts.*

(2) *There exists an object* X *in* T *such that* $H \cong \mathrm{Hom}_{\mathsf{T}}(-, X)$. $\qquad\square$

The proof is in some sense constructive; it shows that any object in T arises as the homotopy colimit of a sequence of morphisms

$$X_0 \xrightarrow{\phi_0} X_1 \xrightarrow{\phi_1} X_2 \xrightarrow{\phi_2} \cdots$$

such that X_0 and the cone of each ϕ_i is a coproduct of objects of the form $\Sigma^n C$, with $n \in \mathbb{Z}$ and C an object from the set of compact objects generating T.

Here are some useful consequences of the Brown representability theorem. The description of compact generation in Proposition 1.47 is another application.

Corollary 2.12. *Each compactly generated triangulated category admits set-indexed products.*

Proof. Given a set of objects $\{X_i\}_{i \in I}$ in such a category T, apply the Brown representability theorem to the functor $Y \mapsto \prod_{i \in I} \mathrm{Hom}_{\mathsf{T}}(Y, X_i)$; check that it is cohomological and takes set-indexed coproducts in T to products in Ab. $\qquad\square$

Corollary 2.13. *Let* $F \colon \mathsf{T} \to \mathsf{U}$ *be an exact functor between triangulated categories and suppose that* T *is compactly generated. Then* F *preserves set-indexed coproducts if and only if* F *admits a right adjoint.*

Proof. If F preserves set-indexed coproducts, then for each object $X \in \mathsf{U}$ the functor $\mathrm{Hom}_{\mathsf{U}}(F-, X)$ is representable by an object $Y \in \mathsf{T}$. Sending X to Y yields a right adjoint of F. The converse is clear, since each left adjoint preserves set-indexed coproducts.   □

2.2.2   Localisation functors

A functor $L \colon \mathsf{C} \to \mathsf{C}$ is called a *localisation functor* if there exists a morphism $\eta \colon \mathrm{Id}_{\mathsf{C}} \to L$ such that the morphism $L\eta \colon L \to L^2$ is invertible and $L\eta = \eta L$. Recall that a morphism $\mu \colon F \to G$ between functors is *invertible* if and only if for each object X the morphism $\mu X \colon FX \to GX$ is an isomorphism. Note that we only require the existence of η; the actual morphism is not part of the definition of L because it is determined by L up to a unique isomorphism $L \to L$.

A functor $\Gamma \colon \mathsf{C} \to \mathsf{C}$ is called *colocalisation functor* if its opposite functor $\Gamma^{\mathrm{op}} \colon \mathsf{C}^{\mathrm{op}} \to \mathsf{C}^{\mathrm{op}}$ is a localisation functor. In this case there is a morphism $\theta \colon \Gamma \to \mathrm{Id}_{\mathsf{C}}$ such that $\theta\Gamma \colon \Gamma^2 \to \Gamma$ is invertible and $\theta\Gamma = \Gamma\theta$.

The following lemma provides an alternative description of a localisation functor. Given any class of morphisms $S \subseteq \mathrm{Mor}\,\mathsf{C}$, the notation $\mathsf{C}[S^{-1}]$ refers to the universal category obtained from C by inverting all morphisms in S; see [32].

Lemma 2.14. *Let $L \colon \mathsf{C} \to \mathsf{C}$ be a functor and $\eta \colon \mathrm{Id}_{\mathsf{C}} \to L$ a morphism. The following conditions are equivalent.*

(1) *The morphism $L\eta \colon L \to L^2$ is invertible and $L\eta = \eta L$.*

(2) *There exists an adjoint pair of functors $F \colon \mathsf{C} \to \mathsf{D}$ and $G \colon \mathsf{D} \to \mathsf{C}$, with F the left adjoint and G the right adjoint, such that G is fully faithful, $L = GF$, and $\eta \colon \mathrm{Id}_{\mathsf{C}} \to GF$ is the adjunction morphism.*

In this case, $F \colon \mathsf{C} \to \mathsf{D}$ induces an equivalence $\mathsf{C}[S^{-1}] \xrightarrow{\sim} \mathsf{D}$, where

$$S = \{\sigma \in \mathrm{Mor}\,\mathsf{C} \mid F\sigma \text{ is invertible}\} = \{\sigma \in \mathrm{Mor}\,\mathsf{C} \mid L\sigma \text{ is invertible}\}.\qquad □$$

Example 2.15. Let A be a commutative ring and denote by $S^{-1}A$ the localisation with respect to a multiplicatively closed subset $S \subseteq A$. Consider the functors $F \colon \mathsf{Mod}\,A \to \mathsf{Mod}\,S^{-1}A$ and $G \colon \mathsf{Mod}\,S^{-1}A \to \mathsf{Mod}\,A$ defined by

$$FM = S^{-1}M = M \otimes_A S^{-1}A \quad and \quad GN = N\,.$$

Then the composite $L = GF$ is a localisation functor.

2.2.3   Acyclic objects and local objects

Let T be a triangulated category which admits set-indexed products and coproducts. We write Σ for the suspension functor on T. Recall that a localising subcategory of T is a full triangulated subcategory that is closed under taking all coproducts. Analogously, a *colocalising subcategory* of T is a full triangulated subcategory that is closed under taking all products.

The *kernel* of an exact functor $F: \mathsf{T} \to \mathsf{T}$ is the full subcategory

$$\operatorname{Ker} F = \{X \in \mathsf{T} \mid FX = 0\},$$

while the *essential image* of F is the full subcategory

$$\operatorname{Im} F = \{X \in \mathsf{T} \mid X \cong FY \text{ for some } Y \text{ in } \mathsf{T}\}.$$

Let $L: \mathsf{T} \to \mathsf{T}$ be an exact localisation functor with morphism $\eta: \operatorname{Id}_\mathsf{T} \to L$ such that $L\eta = \eta L$ is invertible. Note that the kernel of L is a localising subcategory of T, because it coincides with the kernel of a functor that admits a right adjoint by Lemma 2.14. Completing the natural morphism $\eta X: X \to LX$ yields for each object X in T a natural exact *localisation triangle*

$$\Gamma X \longrightarrow X \longrightarrow LX \longrightarrow .$$

This exact triangle gives rise to an exact functor $\Gamma: \mathsf{T} \to \mathsf{T}$ with

$$\operatorname{Ker} L = \operatorname{Im} \Gamma \quad \text{and} \quad \operatorname{Ker} \Gamma = \operatorname{Im} L.$$

The objects in $\operatorname{Ker} L$ are called *L-acyclic* and the objects in $\operatorname{Im} L$ are *L-local*.

The functor Γ is a colocalisation functor, and each colocalisation functor on T arises in this way. This yields a natural bijection between localisation and colocalisation functors on T. Note that $\operatorname{Ker} \Gamma$ is a colocalising subcategory of T.

Given a subcategory C of T, consider full subcategories

$$^{\perp}\mathsf{C} = \{X \in \mathsf{T} \mid \operatorname{Hom}_\mathsf{T}(X, \Sigma^n Y) = 0 \text{ for all } Y \in \mathsf{C} \text{ and } n \in \mathbb{Z}\},$$

$$\mathsf{C}^{\perp} = \{X \in \mathsf{T} \mid \operatorname{Hom}_\mathsf{T}(\Sigma^n Y, X) = 0 \text{ for all } Y \in \mathsf{C} \text{ and } n \in \mathbb{Z}\}.$$

Evidently, $^{\perp}\mathsf{C}$ is a localising subcategory, and $\mathsf{C}^{\perp}$ is a colocalising subcategory.

The next lemma summarises basic facts about localisation and colocalisation.

Proposition 2.16. *Let* T *be a triangulated category and* S *a triangulated subcategory. Then the following are equivalent:*

(1) *There exists a localisation functor* $L: \mathsf{T} \to \mathsf{T}$ *such that* $\operatorname{Ker} L = \mathsf{S}$.

(2) *There exists a colocalisation functor* $\Gamma: \mathsf{T} \to \mathsf{T}$ *such that* $\operatorname{Im} \Gamma = \mathsf{S}$.

In that case both functors are related by a functorial exact triangle

$$\Gamma X \longrightarrow X \longrightarrow LX \longrightarrow .$$

Moreover, there are equalities

$$\mathsf{S}^{\perp} = \operatorname{Im} L = \operatorname{Ker} \Gamma \quad \text{and} \quad {}^{\perp}(\mathsf{S}^{\perp}) = \mathsf{S}.$$

The functor Γ *induces a right adjoint for the inclusion* $\mathsf{S} \to \mathsf{T}$, *and* L *induces a left adjoint for the inclusion* $\mathsf{S}^{\perp} \to \mathsf{T}$. □

Let us mention that L induces an equivalence of categories $\mathsf{T}/\operatorname{Ker} L \xrightarrow{\sim} \operatorname{Im} L$ where $\mathsf{T}/\operatorname{Ker} L$ denotes the Verdier quotient of T with respect to $\operatorname{Ker} L$. This can be deduced from Lemma 2.14, but we do not need this fact.

2.3   The stable module category of a finite group

This lecture consists of an introduction to the cohomology of groups, leading to the definition of the cohomological variety of a kG-module.

2.3.1   Algebra

We begin with the algebraists view of group cohomology. Let k be a commutative ring and G a group. Sooner or later, we shall specialise to the case where k is a field and G is finite, but for the moment that is not necessary.

If M and N are kG-modules, look at

$$\mathrm{Ext}^*_{kG}(N, M).$$

We recall that this is defined as follows. Take a projective resolution of N as a kG-module

$$\cdots \to P_n \to \cdots \to P_1 \to P_0 \to 0$$

with $\mathrm{Coker}(P_1 \to P_0) \cong N$. Apply $\mathrm{Hom}_{kG}(-, M)$ to get

$$0 \to \mathrm{Hom}_{kG}(P_0, M) \to \mathrm{Hom}_{kG}(P_1, M) \to \cdots$$

and the cohomology of this complex in nth place is $\mathrm{Ext}^n_{kG}(N, M)$.

If k is a commutative ring of coefficients and M is a kG-module, we define

$$H^*(G, M) = \mathrm{Ext}^*_{\mathbb{Z}G}(\mathbb{Z}, M) \cong \mathrm{Ext}^*_{kG}(k, M).$$

The isomorphism holds because, if P is a projective resolution of $\mathbb{Z}$ over $\mathbb{Z}G$, then $k \otimes_{\mathbb{Z}} P$ is a projective resolution of k over kG, and adjunction gives an isomorphism of complexes $\mathrm{Hom}_{\mathbb{Z}G}(P, M) \cong \mathrm{Hom}_{kG}(k \otimes_{\mathbb{Z}} P, M)$.

There are two ways to define products in group cohomology, namely *cup products* and *Yoneda products*. When they are both defined, they coincide.

The cup product defines maps

$$H^i(G, M) \otimes_k H^j(G, N) \to H^{i+j}(G, M \otimes_k N)$$

whereas the Yoneda product defines maps

$$\mathrm{Ext}^i_{kG}(M, L) \otimes_k \mathrm{Ext}^j_{kG}(N, M) \to \mathrm{Ext}^{i+j}_{kG}(N, L).$$

These coincide when $L = M = N = k$, giving maps

$$H^i(G, k) \otimes_k H^j(G, k) \to H^{i+j}(G, k).$$

This makes $H^*(G, k)$ into a *graded commutative ring*, meaning that the multiplication satisfies

$$xy = (-1)^{|x||y|} yx.$$

Here, $|x|$ denotes the degree of an element x. Furthermore, cup and Yoneda product give the same module structure for $H^*(G, M)$ as a graded $H^*(G, k)$-module.

It is worth making a comment about graded commutativity versus commutativity. If R is a graded commutative ring and $x \in R$ has odd degree, then

$$x^2 = -x^2$$

and so $2x^2 = 0$, which implies that $(2x)^2 = 0$. Furthermore, again using graded commutativity we see that for all $y \in R$, $(2xy)^2 = \pm(2x)^2 y^2 = 0$. So $2x$ is in the *nil radical* of R, and it follows that modulo the nil radical, x is congruent to $-x$.

The conclusion of this argument is as follows. If R is graded commutative, then R modulo its nil radical is strictly commutative. So it makes sense to talk about the maximal ideal spectrum or the prime ideal spectrum of R.

Theorem 2.17 (Evens [29] (1961)). *Let G be a finite group and k a commutative ring of coefficients. If M is a noetherian k-module, then $H^*(G, M)$ is a noetherian $H^*(G, k)$-module. In particular, if k is a noetherian ring, then so is $H^*(G, k)$.* $\square$

We are interested in the case where k is a field of characteristic p. In this case, Evens' theorem implies that $H^*(G, k)$ is a finitely generated graded commutative k-algebra. If $\operatorname{char}(k)$ is zero or does not divide $|G|$, then there is nothing interesting here. One just gets k in degree zero. More generally, for any commutative ring of coefficients k, $|G|$ annihilates positive degree elements.

The cohomology of elementary abelian groups was described in Section 1.2.2. Here is another interesting example; more are given later on in this section.

Example 2.18. Let $G = M_{11}$, the Mathieu group, and suppose $\operatorname{char}(k) = 2$. Then $H^*(G, k) = k[x, y, z]/(x^2 y + z^2)$ where $|x| = 3$, $|y| = 4$ and $|z| = 5$.

To get further with the commutative algebra of the cohomology ring, we next turn to Quillen's stratification theorem [48, 49]. The first aspect is a characterisation of nilpotent elements, which should remind the reader of Chouinard's theorem.

Theorem 2.19 (Quillen). *Let G be a finite group and k a field of characteristic p. An element of $H^*(G, k)$ is nilpotent if and only if its restriction to every elementary abelian p-subgroup is nilpotent.* $\square$

The next aspect is a determination of the Krull dimension of cohomology.

Theorem 2.20 (Quillen). *The Krull dimension of $H^*(G, k)$ is equal to the p-rank of G, namely the largest r for which $(\mathbb{Z}/p)^r \leq G$.* $\square$

More generally, Quillen described the prime ideal spectrum of $H^*(G, k)$ in terms of the set of elementary abelian subgroups of G, and their conjugations and inclusions. In his paper, he talks in terms of the inhomogeneous maximal ideal spectrum, so we shall begin by doing the same. Since $H^*(G, k)$ modulo its nil radical is a finitely generated commutative k-algebra, Hilbert's Nullstellensatz tells us that the prime ideal spectrum is determined by the maximal ideal spectrum and vice versa.

Theorem 2.21 (Quillen). *Let $V_G = \operatorname{Max} H^*(G, k)$, the spectrum of maximal ideals in $H^*(G, k)$. Then at the level of sets we have*

$$V_G = \varinjlim \operatorname{Max} H^*(E, k). \qquad\qquad \square$$

The limit is over the "Quillen category" whose objects are the elementary abelian p-subgroups of G and whose arrows are group homomorphisms induced by composing conjugations in G and inclusions.

It follows from Propositions 1.36 and 1.37 that $\operatorname{Max} H^*(E, k)$ is an affine space of dimension equal to $\operatorname{rank}(E)$, and this describes $\operatorname{Max} H^*(G, k)$ as layers of affine spaces quotiented by actions of normalisers and glued together.

In the last lecture we shall give a more precise statement of Quillen's stratification theorem, because it is an essential ingredient in the classification of localising subcategories of the stable module category.

Remark 2.22. The homogeneous prime spectrum of a nonstandardly graded ring can be quite confusing. For example, its spectrum can have singularities even though the ring itself is regular in the commutative algebra sense. An explicit example of this is the ring $k[x, y, z]$ with $|x| = 2, |y| = 4, |z| = 6$. The affine open patch corresponding to $y \neq 0$ has coordinate ring generated by $\alpha = x^2/y$, $\beta = xz/y^2$, and $\gamma = z^2/y^3$, with the single relation $\alpha\gamma = \beta^2$. This has a singularity at the origin. This example arises as $H^*(G, k)$ modulo its nil radical with $G = (\mathbb{Z}/p)^3 \rtimes \Sigma_3$ for $p \geq 5$, and k a field of characteristic p.

2.3.2 Topology

Next we examine group cohomology from the point of view of an algebraic topologist. Strictly speaking, this is not necessary for the development of the subject of these lectures, but it would be a bad idea to know about group cohomology only from an algebraic standpoint, as many of the ideas in the subject come from algebraic topology.

Let EG be a contractible space with a free G-action. Such a space always exists, even for a topological group, by a theorem of Milnor. The quotient space by the action of G, denoted $BG = EG/G$ is independent of choice of EG, up to homotopy. This space BG is called the *classifying space* of the group G.

The topologist's definition of group cohomology is the cohomology of the space BG:

$$H^*(G, k) = H^*(BG; k).$$

We reconcile this with the algebraic definition

$$H^*(G, k) = \operatorname{Ext}^*_{\mathbb{Z}G}(\mathbb{Z}, k)$$

as follows: The singular chain complex $C_*(EG)$ has a $\mathbb{Z}$-basis consisting of the singular simplices; in degree n these are the continuous maps $\Delta^n \to EG$ where Δ^n is a standard simplex. Since EG is contractible, this is an acyclic complex except in degree zero, where it augments onto $\mathbb{Z}$. Since the action of G on singular simplices is free, $C_*(EG)$ is a complex of free $\mathbb{Z}G$-modules. This means that it is a *free resolution* of $\mathbb{Z}$ as a $\mathbb{Z}G$-module. So we have

$$
\begin{aligned}
H^*(BG; k) &= H^*(\mathrm{Hom}_{\mathbb{Z}}(C_*(BG), k)) \\
&= H^*(\mathrm{Hom}_{\mathbb{Z}G}(C_*(EG), k)) \\
&\cong \mathrm{Ext}^*_{\mathbb{Z}G}(\mathbb{Z}, k).
\end{aligned}
$$

2.3.3 Examples

Example 2.23. Our first example is an infinite group. Let $G = \mathbb{Z}$, the additive group of integers. We can take $EG = \mathbb{R}$ with G acting by translation. Then $BG = \mathbb{R}/\mathbb{Z} = S^1$, the circle. Thus

$$
H^0(\mathbb{Z}, k) \cong H^1(\mathbb{Z}, k) \cong k, \qquad H^i(\mathbb{Z}, k) = 0 \text{ for } i \geq 2.
$$

Example 2.24. Let $G = \mathbb{Z}/2$, the cyclic group of order two. Then G acts freely on an n-sphere S^n by sending each point to the antipodal point. This is not a contractible space, but if we embed S^n in S^{n+1} equatorially, then it contracts down to the south pole. So taking the union of all the S^n, each embedded equatorially in the next, we get the infinite sphere S^∞ with the weak topology with respect to the subspaces S^n. This is then a contractible space with a free G-action, so we take $EG = S^\infty$. Then the quotient is $BG = \mathbb{R}P^\infty$, the infinite real projective space, again with the weak topology. If k is a field of characteristic 2, then

$$
H^*(G, k) = H^*(\mathbb{R}P^\infty; k) \cong k[x]
$$

with $|x| = 1$. See 1.2.2 for an algebraic perspective on this example and the next.

Example 2.25. For use later in this lecture, we next look at the Klein four group $G = \mathbb{Z}/2 \times \mathbb{Z}/2$. In this case we can take $EG = S^\infty \times S^\infty$ and $BG = \mathbb{R}P^\infty \times \mathbb{R}P^\infty$. If $\mathrm{char}(k) = 2$, then using the Künneth theorem we obtain

$$
H^*(G, k) = H^*(\mathbb{R}P^\infty \times \mathbb{R}P^\infty; k) \cong H^*(\mathbb{R}P^\infty; k) \otimes H^*(\mathbb{R}P^\infty; k) \cong k[x, y]
$$

with $|x| = |y| = 1$.

Example 2.26. Let $G = Q_8$, the quaternions, viewed as the vectors of length 1 in $SU(2) \cong S^3$. Then G acts freely on S^3 by left multiplication. Taking cellular chains $C_*(S^3)$ and splicing in the homology gives us a chain complex

$$
0 \to \mathbb{Z} \to C_3 \to C_2 \to C_1 \to C_0 \to \mathbb{Z} \to 0
$$

where $C_0, \ldots, C_3$ are free $\mathbb{Z}G$-modules.

Form an infinite splice of copies of this complex:

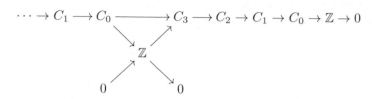

to obtain a free resolution of $\mathbb{Z}$ as a $\mathbb{Z}G$-module.

From the existence of this periodic resolution, we can deduce that for any coefficients k, the cohomology $H^*(Q_8, k)$ is periodic with period 4. The interesting case for us is when k is a field of characteristic 2. In this case the periodicity is given by multiplication by an element $z \in H^4(Q_8, k)$, and we have

$$H^*(Q_8, k)/(z) \cong H^*(S^3/Q_8; k).$$

The same argument proves the following general theorem.

Theorem 2.27. *If G acts freely on S^{n-1}, then $H^*(G, k)$ is periodic with period dividing n.* $\square$

Example 2.28. If $G = \mathbb{Z}/2 \times \mathbb{Z}/2$, then G cannot act freely on any sphere of any dimension, because we computed $H^*(G, k) = k[x, y]$ and this is not periodic.

2.3.4   Cohomological varieties

Definition 2.29. Let G be a finite group and k a field of characteristic p. If M is a finitely generated kG-module, we have a map

$$H^*(G, k) = \operatorname{Ext}^*_{kG}(k, k) \xrightarrow{-\otimes_k M} \operatorname{Ext}^*_{kG}(M, M).$$

Let I_M be the kernel of this map, an ideal of $H^*(G, k)$. Then we define $V_G(M)$ to be the subvariety of $V_G = \operatorname{Max} H^*(G, k)$ determined by this ideal.

The following properties should be compared with the properties of the rank variety $V^r_E(M)$ in Section 1.1.7.

Theorem 2.30. *Let M, N be finitely generated kG-modules.*

(1) $V_G(M)$ *is a closed homogeneous subvariety of V_G.*

(2) $V_G(M) = \{0\}$ *if and only if M is projective.*

(3) $V_G(M \oplus N) = V_G(M) \cup V_G(N)$.

(4) $V_G(M \otimes_k N) = V_G(M) \cap V_G(N)$.

(5) $\dim V_G(M)$ *measures the rate of growth of the minimal projective resolution of M.*

(6) *If $G = E$ is elementary abelian, then there is an isomorphism $V_E \cong \mathbb{A}^r(k)$ taking $V_E(M)$ to $V_E^r(M)$ for every M.*

   Note: for p odd, this isomorphism involves a Frobenius twist. □

The next step is a notion of support for infinitely generated modules.

2.3.5 Rickard idempotents

Let G be a finite group and k a field whose characteristic divides $|G|$. Recall that $\mathsf{StMod}(kG)$ has the same objects as $\mathsf{Mod}(kG)$ but the arrows are module homomorphisms modulo those that factor through a projective module. The following result is the entry point for the application of homotopy theoretic techniques to the modular representation theory of finite groups.

Theorem 2.31. *The stable module category $\mathsf{StMod}(kG)$ is triangulated, with suspension defined by Ω^{-1}. Moreover, it is compactly generated, with compact objects the finite-dimensional modules:*

$$\mathsf{stmod}(kG) \simeq \mathsf{StMod}(kG)^c.$$

Thus, the simple modules are a compact generating set for $\mathsf{StMod}(kG)$; in particular, if G is a p-group, then k is a compact generator for $\mathsf{StMod}(kG)$.

Proof. Since kG is a finite-dimensional and self-injective, this result is a special case of Example 1.48. □

The next result is due to Rickard [53].

Theorem 2.32. *Given a thick subcategory C of $\mathsf{stmod}(kG)$, there exists a functorial triangle in $\mathsf{StMod}(kG)$, unique up to isomorphism,*

$$\Gamma_{\mathsf{C}}(M) \to M \to L_{\mathsf{C}}(M) \to$$

such that $\Gamma_{\mathsf{C}}(M)$ is in the localising subcategory of $\mathsf{StMod}(kG)$ generated by C and there are no maps in $\mathsf{StMod}(kG)$ from C, or equivalently, from $\mathrm{Loc}(\mathsf{C})$ to $L_{\mathsf{C}}(M)$.

Proof. The localizing subcategory $\mathrm{Loc}(\mathsf{C})$ generated by C viewed as a subcategory of $\mathsf{StMod}(kG)$ is obviously compactly generated, and the inclusion

$$\mathrm{Loc}(\mathsf{C}) \subseteq \mathsf{StMod}(kG)$$

preserves coproducts. Brown representability thus yields a right adjoint, which we denote Γ_{C}; see Corollary 2.13. The existence of the functor L, and the remaining assertions, now follow from Proposition 2.16. □

By construction, see Proposition 2.16, the functors Γ_{C} and L_{C} are orthogonal idempotents in the sense that

$$\Gamma_{\mathsf{C}}\Gamma_{\mathsf{C}} \cong \Gamma_{\mathsf{C}}, \qquad L_{\mathsf{C}}L_{\mathsf{C}} \cong L_{\mathsf{C}}, \qquad \Gamma_{\mathsf{C}}L_{\mathsf{C}} \cong L_{\mathsf{C}}\Gamma_{\mathsf{C}} \cong 0.$$

Definition 2.33. Recall that a collection $\mathcal{V}$ of closed homogeneous irreducible subvarieties of V_G is *specialisation closed* if $V \in \mathcal{V}$, $W \subseteq V \Rightarrow W \in \mathcal{V}$.

We write $\mathsf{C}_{\mathcal{V}}$ for the thick subcategory of $\mathsf{stmod}(kG)$ generated by finitely generated modules M such that every irreducible component of $V_G(M)$ is contained in $\mathcal{V}$, and set

$$\Gamma_{\mathcal{V}} = \Gamma_{\mathsf{C}_{\mathcal{V}}} \quad \text{and} \quad L_{\mathcal{V}} = L_{\mathsf{C}_{\mathcal{V}}}.$$

In Section 2.3.7 we prove that the $\mathsf{C}_{\mathcal{V}}$ are the only thick subcategories of $\mathsf{stmod}(kG)$, provided G is a p-group.

2.3.6   Varieties for $\mathsf{StMod}(kG)$

We write $\mathcal{V}_G(k)$ for the set of non-zero closed homogeneous irreducible subvarieties of $V_G = \max H^*(G,k)$, or equivalently the set of non-maximal homogeneous prime ideals in $H^*(G,k)$.

Definition 2.34. Let $V \in \mathcal{V}_G(k)$. Set $\mathcal{V} = \{W \subseteq V\}$ and $\mathcal{W} = \mathcal{V} \setminus \{V\}$, and define

$$\Gamma_V = \Gamma_{\mathcal{V}} L_{\mathcal{W}} = L_{\mathcal{W}} \Gamma_{\mathcal{V}}.$$

Definition 2.35. If M is in $\mathsf{StMod}(kG)$, we define

$$\mathcal{V}_G(M) = \{V \in \mathcal{V}_G(k) \mid \Gamma_V(M) \neq 0\}.$$

The following list of properties should be compared with the properties of $\mathcal{V}_E^r(M)$ from Section 1.1.7.

Theorem 2.36. *Let M be a kG-module.*

(1) $\mathcal{V}_G(M) = \varnothing$ *if and only if M is projective.*

(2) $\mathcal{V}_G(M \oplus N) = \mathcal{V}_G(M) \cup \mathcal{V}_G(N)$, *and more generally*

$$\mathcal{V}_G\Big(\bigoplus_{\alpha} M_{\alpha}\Big) = \bigcup_{\alpha} \mathcal{V}_G(M_{\alpha}).$$

(3) $\mathcal{V}_G(M \otimes_k N) = \mathcal{V}_G(M) \cap \mathcal{V}_G(N)$.

(4) *If M is finitely generated, then* $\mathcal{V}_G(M) = \{V \in \mathcal{V}_G(k) \mid V \subseteq V_G(M)\}$.

(5) *For any subset $\mathcal{V} \subseteq \mathcal{V}_G(k)$ there exists a kG-module M such that $\mathcal{V}_G(M) = \mathcal{V}$.*

(6) *If $G = E$ is elementary abelian, then the isomorphism $V_E \cong \mathbb{A}^r(k)$ of Theorem 2.30 takes $\mathcal{V}_E(M)$ to $\mathcal{V}_E^r(M)$ for every M.* $\square$

Remark 2.37. Observe that when dealing with the stable category $\mathsf{stmod}(kG)$ of finitely generated modules, we used V_G, the spectrum of maximal ideals in $H^*(G,k)$, whereas for the stable category $\mathsf{StMod}(kG)$ of all modules we used $\mathcal{V}_G$, the spectrum of homogeneous prime ideals in $H^*(G,k)$. In the case of a finitely generated module, $V_G(M)$ is determined by $\mathcal{V}_G(M)$, by the theorem above. It would have been possible to use $\mathcal{V}_G$ throughout, but we chose not to, partly for historical reasons. The origin of the use of V_G is Quillen's work [49], and much of the literature on finitely generated modules has been written in this context.

2.3.7 Classification of thick subcategories

We begin with a definition.

Definition 2.38. Write $\text{Thick}(M)$ for the thick subcategory of $\mathsf{stmod}(kG)$ generated by a finitely generated kG-module M.

The following result due to Benson, Carlson, and Rickard [9] is the analogue of Hopkins' theorem 2.8 for the stable module category. For the sake of simplicity we stick to the case of p-groups. For a more general finite group there are some extra technicalities.

Theorem 2.39. *Let G be a p-group, let M be a finitely generated kG-module, and set $\mathcal{V} = \{V \in \mathcal{V}_G(k) \mid V \subseteq \mathcal{V}_G(M)\}$. Then $\mathsf{C}_{\mathcal{V}} = \text{Thick}(M)$.*

Remark 2.40. The hypothesis that G is a p-group comes in as follows: For any kG-module M, if $\underline{\text{Hom}}_{kG}(k, M) = 0$, then $\underline{\text{Hom}}_{kG}(M, M) = 0$, since k is a compact generator for $\mathsf{StMod}(kG)$, by Theorem 2.31, and hence M is projective.

Proof. Let $\mathsf{C}' = \text{Thick}(M)$ and $\mathsf{C} = \mathsf{C}_{\mathcal{V}}$. It is clear that M is in C, so $\mathsf{C}' \subseteq \mathsf{C}$. Consider Rickard triangles associated with these subcategories:

$$\Gamma_{\mathsf{C}} \to \text{Id} \to L_{\mathsf{C}} \to, \qquad \Gamma_{\mathsf{C}'} \to \text{Id} \to L_{\mathsf{C}'} \to .$$

Since $\mathsf{C}' \subseteq \mathsf{C}$, there is an equality $\Gamma_{\mathsf{C}'}\Gamma_{\mathsf{C}} = \Gamma_{\mathsf{C}'}$. Therefore if N is in $\mathsf{StMod}(kG)$, there is an exact triangle

$$\Gamma_{\mathsf{C}'}(N) \to \Gamma_{\mathsf{C}}(N) \to L_{\mathsf{C}'}\Gamma_{\mathsf{C}}(N) \to .$$

So $L_{\mathsf{C}'}\Gamma_{\mathsf{C}}(N)$ is in $\text{Loc}(\mathsf{C})$. This implies that $\mathcal{V}_G(L_{\mathsf{C}'}\Gamma_{\mathsf{C}}(N)) \subseteq \mathcal{V}$. But there are no homomorphisms from M to $L_{\mathsf{C}'}\Gamma_{\mathsf{C}}(N)$, so with $M^* = \text{Hom}_k(M, k)$, one gets

$$0 = \underline{\text{Hom}}_{kG}(M, L_{\mathsf{C}'}\Gamma_{\mathsf{C}}(N)) \cong \underline{\text{Hom}}_{kG}(k, \text{Hom}_k(M, L_{\mathsf{C}'}\Gamma_{\mathsf{C}}(N)))$$
$$\cong \underline{\text{Hom}}_{kG}(k, M^* \otimes_k L_{\mathsf{C}'}\Gamma_{\mathsf{C}}(N))$$

where the first isomorphism is by adjunction while the second one holds because M is finite-dimensional over k. Since G is a p-group, this implies that $M^* \otimes_k L_{\mathsf{C}'}\Gamma_{\mathsf{C}}(N)$ is 0 in the stable module category. Noting that $\mathcal{V}_G(M^*) = \mathcal{V}_G(M)$ (see the exercises to this lecture) the tensor product theorem, 2.36(3), then yields

$$\mathcal{V}_G(M) \cap \mathcal{V}_G(L_{\mathsf{C}'}\Gamma_{\mathsf{C}}(N)) = \varnothing.$$

Hence $\mathcal{V}_G(L_{\mathsf{C}'}\Gamma_{\mathsf{C}}(N)) = \varnothing$, and so $L_{\mathsf{C}'}\Gamma_{\mathsf{C}}(N) = 0$. So $\Gamma_{\mathsf{C}'}(N) \to \Gamma_{\mathsf{C}}(N)$ is an isomorphism for all N. Finally, if N is in C, then $\Gamma_{\mathsf{C}}(N) \to N$ is an isomorphism, so $\Gamma_{\mathsf{C}'}(N) \to N$ is an isomorphism. This implies that N is in C'. $\square$

The result below can be deduced from Theorem 2.39 along the same lines that Theorem 2.10 is deduced from Theorem 2.8.

Corollary 2.41. *Let G be a p-group. The thick subcategories of* stmod(kG) *are in bijection with subsets of $\mathcal{V}_G(k)$ that are closed under specialisation. Under this bijection, a specialisation closed subset $\mathcal{V} \subseteq \mathcal{V}_G(k)$ corresponds to the full subcategory consisting of those finitely generated kG-modules M with the property that every irreducible component of $V_G(M)$ is an element of $\mathcal{V}$.*   □

Remark 2.42. The extra technicality in the case where the finite group G is not a p-group is the following. There is more than one simple kG-module, and we need to impose the extra condition on our thick subcategories that they are closed under tensor product with each simple module, or equivalently under tensor product with all finitely generated kG-modules. This condition is automatically satisfied in the case of a p-group since every finitely generated module has a finite filtration whose filtered quotients are direct sums of copies of k.

For a general finite group G, we say that a thick subcategory of stmod(kG) is *tensor ideal* if this extra condition holds. The proof above goes through for an arbitrary finite group to yield a classification of the tensor ideal thick subcategories of stmod(kG).

The classification of all thick subcategories of stmod(kG) when G is not a p-group is problematic. There is a subvariety of V_G called the *nucleus* which captures the extra complication in this situation. For further details see [6, 7, 9].

2.4   Exercises

In the following exercises A denotes a commutative noetherian ring, G a finite group, and, depending on the context, M is either a complex of A-modules, or a kG-module defined over a field k, usually with char k dividing $|G|$.

(1) Find supp$_{\mathbb{Z}} M$ for a finitely generated abelian group M. What is supp$_{\mathbb{Z}}(\mathbb{Q}/\mathbb{Z})$?

(2) Compute Supp$_A k(\mathfrak{p})$ and supp$_A k(\mathfrak{p})$ for each prime ideal $\mathfrak{p}$.

(3) Verify that the following subcategories of $\mathsf{D} = \mathsf{D}^b(\mathsf{mod}\, A)$ are thick:

$$\{M \in \mathsf{D} \mid M \text{ is perfect}\},$$
$$\{M \in \mathsf{D} \mid M \text{ has finite injective dimension}\},$$
$$\{M \in \mathsf{D} \mid \text{length}_A H^* M \text{ is finite}\}.$$

Think of other interesting properties of complexes, and consider whether they define thick subcategories of D.

(4) Prove that when M is perfect so is $M /\!\!/ \mathfrak{a}$, for any ideal $\mathfrak{a}$ in A.

(5) Recall Hopkins' theorem: if M and N are perfect complexes over A, then supp$_A M \subseteq$ supp$_A N$ implies $M \in \text{Thick}(N)$.

Even when the A-modules $H^*(M)$ and $H^*(N)$ are finitely generated, the hypothesis that M, N are perfect is needed; find relevant examples.

(6) Prove (without recourse to the result of Hopkins) that if M and N are finitely generated $\mathbb{Z}$-modules (or, even bounded complexes of finitely generated $\mathbb{Z}$-modules) with $\operatorname{Supp}_{\mathbb{Z}} M \subseteq \operatorname{Supp}_{\mathbb{Z}} N$, then M is in $\operatorname{Thick}(N)$.

(7) When G is a p-group and M is a finitely generated kG-module, $V_G(M)$ coincides with the subvariety of V_G defined by the annihilator of $H^*(G, M)$ as a module over $H^*(G, k)$. Prove this.

(8) Assume M is a finite-dimensional kG-module. Prove that the following conditions are equivalent:

    (a) $V_G(M) = \{0\}$;

    (b) M has finite projective dimension as a kG-module;

    (c) M is a projective kG-module.

(9) Prove that when M is a finite generated kG-module, there is an equality

$$V_G(\operatorname{Hom}_k(M, k)) = V_G(M).$$

(10) Let $G = \mathbb{Z}/2$ and assume char $k = 2$. Prove that $H^*(G, k) \cong k[x]$, a polynomial algebra over k on a indeterminate x of degree 1.

(11) Let $G = \mathbb{Z}/p$ with $p \geq 3$ and char $k = p$. Prove that $H^*(G, k)$ is the tensor product of an exterior algebra and a polynomial algebra, that is, one has an isomorphism of k-algebras: $H^*(G, k) \cong \Lambda(x) \otimes_k k[y]$, with $|x| = 1$ and $|y| = 2$.

(12) The Künneth isomorphism yields an isomorphism of graded k-algebras

$$H^*(G \times H, k) \cong H^*(G, k) \otimes_k H^*(H, k)$$

where the tensor product on the right is the graded tensor product:

$$(a \otimes b) \cdot (c \otimes d) = (-1)^{|b||c|} ac \otimes bd.$$

Using this isomorphism, compute the cohomology algebra of an elementary abelian p-group of rank r, for any $p \geq 2$.

(13) Prove that $V_G \backslash \{0\}$ is connected for any finite group G.

Hint: Use Quillen's theorem, noting that each non-trivial p-group has a central element of order p.

(14) Describe the conjugacy classes of elementary abelian 2-subgroups of Q_8. Then, using Quillen's theorem, compute the variety of Q_8, when char $k = 2$.

(15) Let $G = D_8$, the dihedral group of order 8 and char $k = 2$. The following series of exercises leads to a description of V_G:

    (i) Prove that there are precisely two conjugacy classes of maximal elementary 2-subgroups of G (call them E_1 and E_2) both of rank 2, and that they contain the centre, $\mathbb{Z}/2$, of G.

(ii) Show that E_i is normal in G, so that $G/E_i \cong \mathbb{Z}/2$.

(iii) Recall that $H^*(E_i, k) \cong k[x, y]$, with $|x| = 1 = |y|$. Prove that the $\mathbb{Z}/2$ action on $k[x, y]$, from (ii), is given by $x \mapsto x$ and $y \mapsto x + y$, and that

$$H^*(E_i, k)^{G/E_i} \cong k[x, (x+y)y].$$

(iv) Use Quillen's theorem to deduce that V_G consists of two affine planes glued along a line:

$$V_G = \mathbb{A}^2(k) \cup_{\mathbb{A}^1(k)} \mathbb{A}^2(k).$$

(16) This exercise describes how to rotate a triangle in $\mathsf{StMod}(kG)$ arising from an exact sequence of kG-modules:

$$0 \to A \to B \to C \to 0.$$

Choose a surjective map $P \to C$ with P a projective module. Write ΩC for the kernel of the composite surjection $P \to C$. Prove that there is an exact sequence of kG-modules

$$0 \to \Omega C \to P \oplus A \to B \to 0.$$

One gets a triangle $\Omega C \to A \to B \to$ in $\mathsf{StMod}(kG)$ as $A \cong P \oplus A$ there.

Similarly, if we embed B into an injective module I and form the dual construction, we obtain an exact sequence

$$0 \to B \to I \oplus C \to \Omega^{-1}A \to 0.$$

This gives a triangle $B \to C \to \Omega^{-1}A \to$ in $\mathsf{StMod}(kG)$.

(17) Let A be an abelian category with enough injectives; for example, the category of modules over some ring. If X, Y are complexes over A such that Y^n is injective for all n and $Y^n = 0$ for $n \ll 0$, then the canonical map

$$\mathrm{Hom}_{\mathsf{K}(\mathsf{A})}(X, Y) \to \mathrm{Hom}_{\mathsf{D}(\mathsf{A})}(X, Y)$$

is an isomorphism. Using this, prove that for any M, N in A one has

$$\mathrm{Hom}_{\mathsf{D}(\mathsf{A})}(M, \Sigma^i N) \cong \mathrm{Ext}^i_{\mathsf{A}}(M, N) \quad \text{for all } i \geq 0.$$

Here M, N are viewed as complexes concentrated in degree zero.

(18) Let T be a compactly generated triangulated category. Show that each object $X \neq 0$ admits a non-zero morphism $C \to X$ from a compact object C.

3 Wednesday

The main goal of Wednesday's lectures is to introduce and develop a theory of support for triangulated categories endowed with an action of a commutative ring. The reason for doing this is that the proofs of the main results, to be presented in Friday's lectures, involve various triangulated categories (of modules over group algebras, and differential graded modules over differential graded rings) and one needs a notion of support applicable to each one of these contexts. In fact, what is required is a more fundamental construction, which underlies support, namely, local cohomology functors. This unifies and extends Grothendieck's local cohomology for commutative rings [34] and the technology of Rickard functors for the stable module category, presented in Section 2.3.5.

3.1 Local cohomology and support

In this lecture we explain the construction and the basic properties of local cohomology functors with respect to the action of a graded-commutative noetherian ring. These functors are then used to define the support for objects in a triangulated category. We follow closely the exposition in [10, §§4–5]. For an introduction to local cohomology for commutative rings, on which much of our development is based, see the original source [34]; see [39] for a more recent exposition, highlighting its connections to various branches of mathematics.

3.1.1 Central ring actions

Let T be a triangulated category admitting set-indexed coproducts. Recall that we write Σ for the suspension on T. For objects X and Y in T, let

$$\mathrm{Hom}^*_{\mathsf{T}}(X, Y) = \bigoplus_{i \in \mathbb{Z}} \mathrm{Hom}_{\mathsf{T}}(X, \Sigma^i Y)$$

be the graded abelian group of morphisms. Set $\mathrm{End}^*_{\mathsf{T}}(X) = \mathrm{Hom}^*_{\mathsf{T}}(X, X)$; this is a graded ring, and $\mathrm{Hom}^*_{\mathsf{T}}(X, Y)$ is a right $\mathrm{End}^*_{\mathsf{T}}(X)$ and left $\mathrm{End}^*_{\mathsf{T}}(Y)$-bimodule.

Let R be a graded-commutative ring; thus R is $\mathbb{Z}$-graded and $rs = (-1)^{|r||s|}sr$ for each pair of homogeneous elements r, s in R. We say that T is R-linear, or that

R *acts* on T, if there is a homomorphism $\phi\colon R \to Z^*(\mathsf{T})$ of graded rings, where $Z^*(\mathsf{T})$ is the *graded centre* of T. Note that $Z^*(\mathsf{T})$ is a graded-commutative ring, where for each $n \in \mathbb{Z}$ the component in degree n is

$$Z^n(\mathsf{T}) = \{\eta\colon \operatorname{Id}_\mathsf{T} \to \Sigma^n \mid \eta\Sigma = (-1)^n \Sigma\eta\}.$$

This yields for each object X a homomorphism $\phi_X\colon R \to \operatorname{End}^*_\mathsf{T}(X)$ of graded rings such that for all objects $X, Y \in \mathsf{T}$ the R-module structures on $\operatorname{Hom}^*_\mathsf{T}(X, Y)$ induced by ϕ_X and ϕ_Y coincide, up to the usual sign rule.

For the rest of this lecture, T will be a compactly generated triangulated category with set-indexed coproducts, and R a graded-commutative noetherian ring acting on T.

3.1.2 Local cohomology functors

We write $\operatorname{Spec} R$ for the set of homogeneous prime ideals of R. Fix $\mathfrak{p} \in \operatorname{Spec} R$ and let M be a graded R-module. The homogeneous localisation of M at $\mathfrak{p}$ is denoted by $M_\mathfrak{p}$ and M is called $\mathfrak{p}$-*local* when the natural map $M \to M_\mathfrak{p}$ is bijective.

Given a homogeneous ideal $\mathfrak{a}$ in R, we set

$$\mathcal{V}(\mathfrak{a}) = \{\mathfrak{p} \in \operatorname{Spec} R \mid \mathfrak{p} \supseteq \mathfrak{a}\}.$$

A graded R-module M is $\mathfrak{a}$-*torsion* if each element of M is annihilated by a power of $\mathfrak{a}$; equivalently, if $M_\mathfrak{p} = 0$ for all $\mathfrak{p} \in \operatorname{Spec} R \setminus \mathcal{V}(\mathfrak{a})$.

The *specialisation closure* of a subset $\mathcal{U}$ of $\operatorname{Spec} R$ is the set

$$\operatorname{cl}\mathcal{U} = \{\mathfrak{p} \in \operatorname{Spec} R \mid \text{there exists } \mathfrak{q} \in \mathcal{U} \text{ with } \mathfrak{q} \subseteq \mathfrak{p}\}.$$

The subset $\mathcal{U}$ is *specialisation closed* if $\operatorname{cl}\mathcal{U} = \mathcal{U}$; equivalently, if $\mathcal{U}$ is a union of Zariski closed subsets of $\operatorname{Spec} R$. For each specialisation closed subset $\mathcal{V}$ of $\operatorname{Spec} R$, we define the full subcategory of T of $\mathcal{V}$-*torsion objects* as follows:

$$\mathsf{T}_\mathcal{V} = \{X \in \mathsf{T} \mid \operatorname{Hom}^*_\mathsf{T}(C, X)_\mathfrak{p} = 0 \text{ for all } C \in \mathsf{T}^c, \mathfrak{p} \in \operatorname{Spec} R \setminus \mathcal{V}\}.$$

This is a localising subcategory and there exists a localisation functor $L_\mathcal{V}\colon \mathsf{T} \to \mathsf{T}$ such that $\operatorname{Ker} L_\mathcal{V} = \mathsf{T}_\mathcal{V}$; see [10, Lemma 4.3, Proposition 4.5].

By Proposition 2.16, the localisation functor $L_\mathcal{V}$ induces a colocalisation functor on T, which we denote $\Gamma_\mathcal{V}$, and call the *local cohomology functor* with respect to $\mathcal{V}$. For each object X in T there is then an exact localisation triangle

$$\Gamma_\mathcal{V} X \longrightarrow X \longrightarrow L_\mathcal{V} X \longrightarrow . \tag{3.1}$$

For each $\mathfrak{p}$ in $\operatorname{Spec} R$ and each object X in T set

$$X_\mathfrak{p} = L_{\mathcal{Z}(\mathfrak{p})} X, \quad \text{where } \mathcal{Z}(\mathfrak{p}) = \{\mathfrak{q} \in \operatorname{Spec} R \mid \mathfrak{q} \nsubseteq \mathfrak{p}\}.$$

Note that $\mathcal{Z}(\mathfrak{p})$ is the subset $\operatorname{Spec} R \setminus \operatorname{Spec} R_{\mathfrak{p}}$. This is one justification for the notation $X_{\mathfrak{p}}$. A more compelling one is [10, Theorem 4.7]: the adjunction morphism $X \to X_{\mathfrak{p}}$ induces for any compact object C an isomorphism of R-modules

$$\operatorname{Hom}_{\mathsf{T}}^*(C, X)_{\mathfrak{p}} \xrightarrow{\cong} \operatorname{Hom}_{\mathsf{T}}^*(C, X_{\mathfrak{p}}).$$

We say X is $\mathfrak{p}$-*local* if the adjunction morphism $X \to X_{\mathfrak{p}}$ is an isomorphism; this is equivalent to the condition that there exists *some* isomorphism $X \cong X_{\mathfrak{p}}$ in T.

Consider the exact functor $\Gamma_{\mathfrak{p}} \colon \mathsf{T} \to \mathsf{T}$ obtained by setting

$$\Gamma_{\mathfrak{p}} X = \Gamma_{\mathcal{V}(\mathfrak{p})}(X_{\mathfrak{p}}) \quad \text{for each object } X \text{ in } \mathsf{T},$$

and let $\Gamma_{\mathfrak{p}} \mathsf{T}$ denote its essential image. One has a natural isomorphism $\Gamma_{\mathfrak{p}}^2 \cong \Gamma_{\mathfrak{p}}$, and an object X from T is in $\Gamma_{\mathfrak{p}} \mathsf{T}$ if and only if the R-module $\operatorname{Hom}_{\mathsf{T}}^*(C, X)$ is $\mathfrak{p}$-local and $\mathfrak{p}$-torsion for every compact object C; see [10, Corollary 4.10].

3.1.3 Support

The *support* of an object X in T is by definition the set

$$\operatorname{supp}_R X = \{\mathfrak{p} \in \operatorname{Spec} R \mid \Gamma_{\mathfrak{p}} X \neq 0\}.$$

The result below, see [10, Theorem 5.15], collects basic properties of support; it is in fact, an axiomatic characterization. It is convenient to set for each X in T

$$\operatorname{supp}_R H^*(X) = \bigcup_{C \in \mathsf{I}^c} \operatorname{supp}_R \operatorname{Hom}_{\mathsf{T}}^*(C, X).$$

Note that the support on the right-hand side refers to that defined in Section A.5.

Theorem 3.2. *The assignment that sends each object X in T to the subset $\operatorname{supp}_R X$ of $\operatorname{Spec} R$ has the following properties:*

(1) Cohomology: *For each object X in T one has*

$$\operatorname{cl}(\operatorname{supp}_R X) = \operatorname{cl}(\operatorname{supp}_R H^*(X)).$$

(2) Orthogonality: *For objects X and Y in T, one has that*

$$\operatorname{cl}(\operatorname{supp}_R X) \cap \operatorname{supp}_R Y = \varnothing \quad \text{implies} \quad \operatorname{Hom}_{\mathsf{T}}(X, Y) = 0.$$

(3) Exactness: *For every exact triangle $X \to Y \to Z \to$ in T, one has*

$$\operatorname{supp}_R Y \subseteq \operatorname{supp}_R X \cup \operatorname{supp}_R Z.$$

(4) Separation: *For any specialisation closed subset $\mathcal{V}$ of $\operatorname{Spec} R$ and object X in T, there exists an exact triangle $X' \to X \to X'' \to$ in T such that*

$$\operatorname{supp}_R X' \subseteq \mathcal{V} \quad \text{and} \quad \operatorname{supp}_R X'' \subseteq \operatorname{Spec} R \setminus \mathcal{V}.$$

Any other assignment having these properties coincides with $\operatorname{supp}_R(-)$. □

3.2   Koszul objects and support

In the last lecture we defined a notion of support for triangulated categories, based on local cohomology functors. In this one we describe some methods for computing it, in terms of the support, in the sense of Appendix A., of the cohomology.

Let R be a graded commutative noetherian ring and T an R-linear triangulated category; see Section 3.1.1. To each specialisation closed subset $\mathcal{V}$ of $\operatorname{Spec} R$ we associated an exact localisation triangle

$$\Gamma_{\mathcal{V}} X \to X \to L_{\mathcal{V}} X \to .$$

One should think of $\Gamma_{\mathcal{V}} X$ as the part of X supported on $\mathcal{V}$ and $L_{\mathcal{V}} X$ as the part supported on $\operatorname{Spec} R \setminus \mathcal{V}$; this will be clarified in this lecture. This point of view should at least make the following statement plausible.

Lemma 3.3. *If $\mathcal{V} \subseteq \mathcal{W}$ are specialisation closed sets, then for each X in T there are natural isomorphisms*

$$\Gamma_{\mathcal{V}} \Gamma_{\mathcal{W}} X \xrightarrow{\cong} \Gamma_{\mathcal{V}} X \xleftarrow{\cong} \Gamma_{\mathcal{W}} \Gamma_{\mathcal{V}} X$$
$$\Gamma_{\mathcal{V}} L_{\mathcal{W}} X = 0 = L_{\mathcal{W}} \Gamma_{\mathcal{V}} X.$$

Proof. The hypothesis implies that $\mathcal{V}$-torsion objects are $\mathcal{W}$-torsion: $\mathsf{T}_{\mathcal{V}} \subseteq \mathsf{T}_{\mathcal{W}}$. Thus, essentially by definition, the natural map $\Gamma_{\mathcal{W}} \Gamma_{\mathcal{V}} X \to \Gamma_{\mathcal{V}} X$ is an isomorphism; equivalently, $L_{\mathcal{W}} \Gamma_{\mathcal{V}} X = 0$, as follows by considering the localisation triangle (3.1) defined by $\mathcal{W}$ and the object $\Gamma_{\mathcal{V}} X$. This settles half the claims.

As to the rest: It is easy to verify that the composition $\Gamma_{\mathcal{V}} \Gamma_{\mathcal{W}}$ is a right adjoint to the inclusion $\mathsf{T}_{\mathcal{V}} \subseteq \mathsf{T}$, and hence isomorphic to $\Gamma_{\mathcal{V}}$, by the uniqueness of adjoints. Applying $\Gamma_{\mathcal{V}}$ to the localisation triangle (3.1) defined by $\mathcal{W}$, it then follows that $\Gamma_{\mathcal{V}} L_{\mathcal{W}} X = 0$.   □

Even when $\mathcal{V}$ is not contained in $\mathcal{W}$, the local cohomology functors interact in a predictable, and useful, way; see [10, Proposition 6.1]. One consequence of these results is worth recording:

Theorem 3.4. *Fix a point $\mathfrak{p}$ in $\operatorname{Spec} R$. For any specialisation closed subsets $\mathcal{V}$ and $\mathcal{W}$ of $\operatorname{Spec} R$ such that $\mathcal{V} \setminus \mathcal{W} = \{\mathfrak{p}\}$ there are natural isomorphisms*

$$L_{\mathcal{W}} \Gamma_{\mathcal{V}} \cong \Gamma_{\mathfrak{p}} \cong \Gamma_{\mathcal{V}} L_{\mathcal{W}} .   □$$

This may be viewed as an expression of the local nature of the functor $\Gamma_{\mathfrak{p}}$.

3.2.1   Computing support

Let X be an object in T. For any compact object C in T, we write $H_C^*(X)$ for $\operatorname{Hom}_{\mathsf{T}}^*(C, X)$, and think of it as the *cohomology of X with respect to C*.

For any subset $\mathcal{U}$ of $\operatorname{Spec} R$, we write $\min \mathcal{U}$ for the prime ideals which are minimal, with respect to inclusion, in the set $\mathcal{U}$, and for any R-module M, set

$$\min_R M = \min(\operatorname{Supp}_R M).$$

This set coincides with $\min(\operatorname{supp}_R M)$, for $\operatorname{Supp}_R M$ is the specialisation closure of $\operatorname{supp}_R M$; see Lemma A.15. The result below, which is [10, Theorem 5.2], is often useful for computing supports.

Theorem 3.5. *For each X in T there is an equality*

$$\operatorname{supp}_R X = \bigcup_{C \in \mathsf{T}^c} \min_R H_C^*(X).$$

In particular, $\operatorname{supp}_R X = \varnothing$ if and only if $X = 0$. $\qquad\square$

We omit the proof, and will focus on its consequences, one of which is:

Theorem 3.6. *For each specialisation closed subset $\mathcal{V}$ of $\operatorname{Spec} R$ there are equalities*

$$\operatorname{supp}_R \Gamma_{\mathcal{V}} X = \operatorname{supp}_R X \cap \mathcal{V},$$
$$\operatorname{supp}_R L_{\mathcal{V}} X = \operatorname{supp}_R X \cap (\operatorname{Spec} R \setminus \mathcal{V}).$$

In particular, $\operatorname{supp}_R X = \operatorname{supp}_R \Gamma_{\mathcal{V}} X \sqcup \operatorname{supp}_R L_{\mathcal{V}} X$.

Proof. To begin with, we verify the

Claim: $\operatorname{supp}_R \Gamma_{\mathcal{V}} X \subseteq \mathcal{V}$ and $\operatorname{supp}_R L_{\mathcal{V}} X \subseteq \operatorname{Spec} R \setminus \mathcal{V}$.

Indeed, for each $\mathfrak{p}$ in $\mathcal{V}$, the construction of $\Gamma_{\mathfrak{p}}$ gives the first equality below:

$$\Gamma_{\mathfrak{p}}(L_{\mathcal{V}} X) = L_{\mathcal{Z}(\mathfrak{p})} \Gamma_{\mathcal{V}(\mathfrak{p})} L_{\mathcal{V}} X = 0,$$

while the second equality holds by Lemma 3.3 since $\mathcal{V}(\mathfrak{p}) \subseteq \mathcal{V}$. Thus $\operatorname{supp}_R L_{\mathcal{V}} X$ is contained in $\operatorname{Spec} R \setminus \mathcal{V}$. On the other hand, since the image of $\Gamma_{\mathcal{V}}$ is in $\mathsf{T}_{\mathcal{V}}$, for each compact object C one gets the second inclusion below:

$$\min_R H_C^*(\Gamma_{\mathcal{V}} X) \subseteq \operatorname{supp}_R H_C^*(\Gamma_{\mathcal{V}} X) \subseteq \mathcal{V}.$$

Thus Theorem 3.5 implies $\operatorname{supp}_R \Gamma_{\mathcal{V}} X \subseteq \mathcal{V}$. This justifies the claim.

Now, since $\Gamma_{\mathfrak{p}}$ is an exact functor, for any $\mathfrak{p} \in \operatorname{Spec} R$, from the localisation triangle (3.1) one gets inclusions

$$\operatorname{supp}_R \Gamma_{\mathcal{V}} X \subseteq \operatorname{supp}_R X \cup \operatorname{supp}_R L_{\mathcal{V}} X,$$
$$\operatorname{supp}_R X \subseteq \operatorname{supp}_R \Gamma_{\mathcal{V}} X \cup \operatorname{supp}_R L_{\mathcal{V}} X.$$

Combining the first inclusion with those in the claim yields:

$$\begin{aligned}
\operatorname{supp}_R \Gamma_{\mathcal{V}} X &\subseteq (\operatorname{supp}_R X \cup \operatorname{supp}_R L_{\mathcal{V}} X) \cap \mathcal{V} \\
&= (\operatorname{supp}_R X \cap \mathcal{V}) \cup (\operatorname{supp}_R L_{\mathcal{V}} X \cap \mathcal{V}) \\
&= \operatorname{supp}_R X \cap \mathcal{V}.
\end{aligned}$$

In the same vein, an inclusion $\operatorname{supp}_R L_{\mathcal{V}} X \subseteq \operatorname{supp}_R X \cap (\operatorname{Spec} R \setminus \mathcal{V})$ holds. The desired equalities follow, as $\operatorname{supp}_R X \subseteq \operatorname{supp}_R \Gamma_{\mathcal{V}} X \cup \operatorname{supp}_R L_{\mathcal{V}} X$. $\qquad\square$

Corollary 3.7. *Let $\mathcal{V}$ be a specialisation closed subset of* $\operatorname{Spec} R$.

(1) $\operatorname{supp}_R X \subseteq \mathcal{V} \iff X \in \mathsf{T}_{\mathcal{V}} \iff \Gamma_{\mathcal{V}} X \xrightarrow{\cong} X \iff L_{\mathcal{V}} X = 0$.

(2) $\mathcal{V} \cap \operatorname{supp}_R X = \varnothing \iff X \xrightarrow{\cong} L_{\mathcal{V}} X \iff \Gamma_{\mathcal{V}} X = 0$.

Proof. It follows from the preceding result that $\operatorname{supp}_R X \subseteq \mathcal{V}$ holds if and only if $\operatorname{supp}_R L_{\mathcal{V}} X = 0$, that is to say, if and only if $L_{\mathcal{V}} X = 0$; see Theorem 3.5. Now (1) follows from the localisation triangle (3.1). A similar argument works for (2). $\quad\square$

Corollary 3.8. *If* $\operatorname{cl}(\operatorname{supp}_R X) \cap \operatorname{supp}_R Y = \varnothing$, *then* $\operatorname{Hom}_{\mathsf{T}}^*(X, Y) = 0$.

Proof. For $\mathcal{V} = \operatorname{cl}(\operatorname{supp}_R X)$ the previous corollary yields isomorphisms

$$\Gamma_{\mathcal{V}} X \xrightarrow{\cong} X \quad \text{and} \quad Y \xrightarrow{\cong} L_{\mathcal{V}} Y.$$

Hence $\operatorname{Hom}_{\mathsf{T}}^*(X, Y) \cong \operatorname{Hom}_{\mathsf{T}}^*(\Gamma_{\mathcal{V}} X, L_{\mathcal{V}} Y) = 0$; the equality holds because there are no non-zero morphisms from $\mathcal{V}$-torsion to $\mathcal{V}$-local objects, by Proposition 2.16. $\quad\square$

Corollary 3.9. *For an object* $X \neq 0$ *in* T *and* $\mathfrak{p} \in \operatorname{Spec} R$ *the following are equivalent.*

(1) $\Gamma_{\mathfrak{p}} X \cong X$.

(2) $\operatorname{supp}_R X = \{\mathfrak{p}\}$.

(3) *The R-module $H_C^*(X)$ is $\mathfrak{p}$-local and $\mathfrak{p}$-torsion for each $C \in \mathsf{T}^c$.*

Proof. (1) $\implies$ (3) For each compact object C the R-module $H_C^*(\Gamma_{\mathfrak{p}} X)$ is $\mathfrak{p}$-local because there are isomorphisms of R-modules

$$H_C^*(\Gamma_{\mathfrak{p}} X) = H_C^*(L_{\mathcal{Z}(\mathfrak{p})} \Gamma_{\mathcal{V}(\mathfrak{p})} X) \cong H_C^*(\Gamma_{\mathcal{V}(\mathfrak{p})} X)_{\mathfrak{p}}.$$

It is also $\mathfrak{p}$-torsion, for it is localisation of the $\mathfrak{p}$-torsion module $H_C^*(\Gamma_{\mathcal{V}(\mathfrak{p})} X)$.

(3) $\implies$ (2) follows from Theorem 3.5, and Lemma A.17.

(2) $\implies$ (1) When $\operatorname{supp}_R X = \{\mathfrak{p}\}$ holds, so does $\operatorname{supp}_R \Gamma_{\mathcal{V}(\mathfrak{p})} X = \{\mathfrak{p}\}$, by Theorem 3.6. Corollary 3.7 then yields isomorphisms

$$X \xleftarrow{\cong} \Gamma_{\mathcal{V}(\mathfrak{p})} X \xrightarrow{\cong} L_{\mathcal{Z}(\mathfrak{p})} \Gamma_{\mathcal{V}(\mathfrak{p})} X = \Gamma_{\mathfrak{p}} X.$$

For the second isomorphism, note that $\mathcal{Z}(\mathfrak{p}) \cap \{\mathfrak{p}\} = \varnothing$. $\quad\square$

3.2.2 Koszul objects

Let C be an object in T and $r \in R^d$, a homogeneous element of degree d. Since T is R-linear, r induces a morphism $C \xrightarrow{r} \Sigma^d C$; completing it to an exact triangle yields an object denoted $C /\!\!/ r$:

$$C \xrightarrow{r} \Sigma^d C \to C /\!\!/ r \to \Sigma C \to .$$

Note that $C/\!\!/ r$ is well defined, but only up to non-unique isomorphism. We call it the *Koszul object* on r. For each X in T, the triangle above induces an exact sequence of graded R-modules:

$$\to H^*_C(X)[-d-1] \xrightarrow{r} H^*_C(X)[-1] \to H^*_{C/\!\!/ r}(X) \to H^*_C(X)[-d] \xrightarrow{r} H^*_C(X) \to \ .$$

This translates to an exact sequence of graded R-modules:

$$0 \longrightarrow \frac{H^*_C(X)}{r H^*_C(X)}[-1] \longrightarrow H^*_{C/\!\!/ r}(X) \longrightarrow (0 :_{H^*_C(X)} r) \longrightarrow 0. \qquad (3.10)$$

Compare this with the exact sequence appearing in the proof of Proposition 2.5.

Given a sequence $\boldsymbol{r} = r_1, \ldots, r_n$ we denote $C/\!\!/ \boldsymbol{r}$ the object obtained by iterated Koszul construction. To be precise, $C/\!\!/ \boldsymbol{r} = C_n$ where

$$C_0 = C \quad \text{and} \quad C_i = C_{i-1}/\!\!/ r_i \quad \text{for } i \geq 1.$$

Finally, given an ideal $\mathfrak{a}$ in R, we write $C/\!\!/ \mathfrak{a}$ for any Koszul object $C/\!\!/ \boldsymbol{r}$, where $\boldsymbol{r}$ is a finite generating set for the ideal $\mathfrak{a}$.

The following result is a consequence of [12, Proposition 2.11(2)]. It shows the (in)dependence of the Koszul object on a generating set for the ideal defining it. For tensor triangulated categories, one has a more precise result; see [36, Lemma 6.0.9].

Lemma 3.11. *If $\mathfrak{a}$ and $\mathfrak{b}$ are ideals in R such that $\mathcal{V}(\mathfrak{a}) \subseteq \mathcal{V}(\mathfrak{b})$, then*

$$\mathrm{Loc}_\mathsf{T}(C/\!\!/ \mathfrak{a}) \subseteq \mathrm{Loc}_\mathsf{T}(C/\!\!/ \mathfrak{b}). \qquad \qquad \square$$

The next result records the basic calculations concerning cohomology of Koszul objects. We give only a sketch of the argument, referring the reader to [10, Lemma 5.11] for details.

Proposition 3.12. *Fix objects C and X in T and a point $\mathfrak{p} \in \mathrm{Spec}\, R$.*

(1) *There exists an integer $s \geq 0$, independent of C and X, such that*

$$\mathfrak{p}^s \mathrm{Hom}^*_\mathsf{T}(C/\!\!/ \mathfrak{p}, X) = 0 = \mathfrak{p}^s \mathrm{Hom}^*_\mathsf{T}(X, C/\!\!/ \mathfrak{p}).$$

In particular, $\mathrm{Hom}^*_\mathsf{T}(C/\!\!/ \mathfrak{p}, X)$ and $\mathrm{Hom}^*_\mathsf{T}(X, C/\!\!/ \mathfrak{p})$ are $\mathfrak{p}$-torsion.

(2) *$C/\!\!/ \mathfrak{p}$ is in $\mathsf{T}_{\mathcal{V}(\mathfrak{p})}$.*

(3) *$\mathrm{Hom}^*_\mathsf{T}(C, X) = 0$ implies $\mathrm{Hom}^*_\mathsf{T}(C/\!\!/ \mathfrak{p}, X) = 0$. The converse holds if the $R_\mathfrak{p}$-module $\mathrm{Hom}^*_\mathsf{T}(C, X)$ is either $\mathfrak{p}$-torsion, or $\mathfrak{p}$-local and finitely generated.*

(4) *When C is compact, there is an isomorphism of R-modules:*

$$\mathrm{Hom}^*_\mathsf{T}(C/\!\!/ \mathfrak{p}, \Gamma_\mathfrak{p} X) \cong \mathrm{Hom}^*_\mathsf{T}(C/\!\!/ \mathfrak{p}, X)_\mathfrak{p}.$$

Sketch of a proof. Statements (1) and (3) follow from (3.10) and an induction on the number of generators for $\mathfrak{p}$. Then (2) is a consequence of (1), while (4) is justified by the isomorphisms

$$\operatorname{Hom}_{\mathsf{T}}^*(C/\!\!/\mathfrak{p}, \Gamma_{\mathcal{V}(\mathfrak{p})}X_\mathfrak{p}) \cong \operatorname{Hom}_{\mathsf{T}}^*(C/\!\!/\mathfrak{p}, X_\mathfrak{p}) \cong \operatorname{Hom}_{\mathsf{T}}^*(C/\!\!/\mathfrak{p}, X)_\mathfrak{p}$$

where the first one follows from (2) and the second holds as $C/\!\!/\mathfrak{p}$ is compact. $\square$

Using Koszul objects one can establish a more precise (and economical) version of Theorem 3.5.

Theorem 3.13. *Let* $\mathsf{G} \subset \mathsf{T}$ *be a compact generating set and* X *an object in* T. *Then*

$$\mathfrak{p} \in \operatorname{supp}_R X \iff H_{C/\!\!/\mathfrak{p}}^*(X) \neq 0 \text{ for some } C \in \mathsf{G}.$$

In particular, there is an equality

$$\operatorname{supp}_R X = \bigcup_{\substack{\mathfrak{p} \in \operatorname{Spec} R \\ C \in \mathsf{G}}} \min_R H_{C/\!\!/\mathfrak{p}}^*(X).$$

Proof. Fix $\mathfrak{p}$ in $\operatorname{Spec} R$. The following equivalences hold:

$$
\begin{aligned}
\Gamma_\mathfrak{p} X \neq 0 &\iff H_C^*(\Gamma_\mathfrak{p} X) \neq 0 && \text{for some } C \in \mathsf{G} \\
&\iff H_{C/\!\!/\mathfrak{p}}^*(\Gamma_\mathfrak{p} X) \neq 0 && \text{by Proposition 3.12(3)} \\
&\iff H_{C/\!\!/\mathfrak{p}}^*(X)_\mathfrak{p} \neq 0 && \text{by Proposition 3.12(4)} \\
&\iff \mathfrak{p} \in \min_R H_{C/\!\!/\mathfrak{p}}^*(X) && \text{by Proposition 3.12(1).}
\end{aligned}
$$

In the second step we have used the fact that $\operatorname{Hom}_{\mathsf{T}}(C, \Gamma_\mathfrak{p} X)$ is $\mathfrak{p}$-local and $\mathfrak{p}$-torsion, by Corollary 3.9. $\square$

3.3 The homotopy category of injectives

In this lecture, we explain how to enlarge the stable module category $\operatorname{StMod}(kG)$ slightly to a category $\mathsf{K}(\operatorname{Inj} kG)$. The effect of this on the variety theory is that it puts back the "missing origin" in the collection of subvarieties of V_G. Most of the material in this lecture is taken from the paper of Benson and Krause [14].

3.3.1 The stable module category and Tate cohomology

Let G be a finite group and k be a field of characteristic p dividing $|G|$. Recall from Section 1.1 that the module category $\operatorname{Mod}(kG)$ has as its objects the kG-modules, and arrows the module homomorphisms. The stable module category $\operatorname{StMod}(kG)$ has the same objects, but its arrows are given by

$$\underline{\operatorname{Hom}}_{kG}(M, N) = \operatorname{Hom}_{kG}(M, N)/P \operatorname{Hom}_{kG}(M, N)$$

where $P\operatorname{Hom}_{kG}(M,N)$ is the subspace of homomorphisms that factor through some projective module. The categories $\operatorname{mod}(kG)$ and $\operatorname{stmod}(kG)$ are the full subcategories of finitely generated modules. The categories $\operatorname{Mod}(kG)$, $\operatorname{mod}(kG)$ are abelian categories while $\operatorname{StMod}(kG)$, $\operatorname{stmod}(kG)$ are triangulated categories with shift Ω^{-1}.

The endomorphisms of k in $\operatorname{StMod}(kG)$ form a graded commutative ring called the *Tate cohomology ring* whose degree n part is

$$\hat{H}^n(G,k) = \underline{\operatorname{Hom}}_{kG}(\Omega^n k, k) \cong \underline{\operatorname{Hom}}_{kG}(\Omega^{n+m}k, \Omega^m k).$$

The multiplication is "shift and compose": if $x \in \hat{H}^m(G,k)$ and $y \in \hat{H}^n(G,k)$, then we choose corresponding homomorphisms $\Omega^m k \to k$ and $\Omega^{n+m}k \to \Omega^m k$ and compose them to obtain a representative for the product xy.

If $n \geq 0$, then $\hat{H}^n(G,k) \cong H^n(G,k)$, while for $n < 0$ *Tate duality* implies that $\hat{H}^n(G,k)$ is the vector space dual of $H^{-n-1}(G,k)$.

The result below identifies groups for which the Tate cohomology ring is noetherian; see [10, Lemma 10.1] for a proof.

Theorem 3.14. *If the Sylow p-subgroups of G are cyclic, then $\hat{H}^*(G,k)$ is a localisation of $H^*(G,k)$ given by inverting any positive-degree non-nilpotent element. In this case $\hat{H}^*(G,k)$ is noetherian.*

If the Sylow p-subgroups of G are not cyclic, then every element of negative degree in $\hat{H}^*(G,k)$ is nilpotent. In this case $\hat{H}^*(G,k)$ is not noetherian. □

Next we describe another view of Tate cohomology. A *Tate resolution* of a module M is obtained by splicing together a projective and an injective resolution:

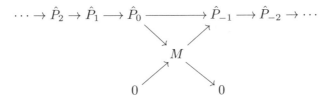

If N is another module, then the Tate Ext group $\widehat{\operatorname{Ext}}_{kG}^n(M,N)$ is the cohomology of the cochain complex $\operatorname{Hom}_{kG}(\hat{P}_*,N)$. In particular the Tate cohomology is given by $\hat{H}^n(G,k) = \widehat{\operatorname{Ext}}_{kG}^n(k,k)$.

3.3.2 The derived category of kG-modules

One of the problems with the stable module category $\operatorname{StMod}(kG)$ is that the graded endomorphism ring of the trivial module is the Tate cohomology ring $\hat{H}^*(G,k)$, which is usually not noetherian. So for example it is a hassle to deal with injective resolutions over this ring. We can try to cure this by using the ordinary cohomology ring, but then the maximal ideal of positive degree elements needs special treatment, and keeping track of this is again a hassle.

We can try to solve this problem by moving to the derived category, but this creates new problems. Let us briefly recall the construction of the derived category. The category of cochain complexes $\mathsf{C}(\mathsf{Mod}\,kG)$ has as objects the complexes of kG-modules and as arrows the degree preserving maps of complexes. The homotopy category of cochain complexes $\mathsf{K}(\mathsf{Mod}\,kG)$ has the same objects, but the arrows are the homotopy classes of maps of complexes. Finally, the derived category $\mathsf{D}(\mathsf{Mod}\,kG)$ has the same objects, but the arrows are obtained by adjoining inverses to the quasi-isomorphisms in $\mathsf{K}(\mathsf{Mod}\,kG)$; recall that a map of cochain complexes is a *quasi-isomorphism* if the induced map between the cohomologies of the complexes is an isomorphism.

Note that $\mathsf{C}(\mathsf{Mod}\,kG)$ is an abelian category while the categories $\mathsf{K}(\mathsf{Mod}\,kG)$ and $\mathsf{D}(\mathsf{Mod}\,kG)$ are triangulated.

If M and N are modules, made into complexes whose only non-zero terms are in degree zero, then the space of degree n homomorphisms in the derived category from M to N is isomorphic to $\mathrm{Ext}^n_{kG}(M,N)$. In particular, the graded endomorphism ring of k is $H^*(G,k)$.

3.3.3   Problems with the derived category

The first problem we have with $\mathsf{D}(\mathsf{Mod}\,kG)$ is that a finitely generated non-projective module M regarded as a complex concentrated in a single degree is not compact. In other words, $\mathrm{Hom}_{kG}(M,-)$ does not distribute over direct sums. So for example

$$\bigoplus_i H^*(G,X_i) \to H^*(G,\bigoplus_i X_i)$$

is not an isomorphism in general. In fact, the compact objects are the *perfect complexes*, namely the complexes isomorphic to bounded complexes of finitely generated projective modules.

Another problem is that there are not a lot of localising subcategories, so it is unlikely to help us directly to classify the localising subcategories of $\mathsf{StMod}(kG)$.

Definition 3.15. We write $\mathsf{Loc}_\mathsf{T}(\mathsf{C})$ for the smallest localising category of a triangulated category T containing a subcategory (or collection of objects) C.

Theorem 3.16. *If G is a p-group, the only localising subcategories of $\mathsf{D}(\mathsf{Mod}\,kG)$ are zero and the entire category.*

Proof. If X is a non-zero object, we claim that $\mathsf{Loc}(X)$ is the whole category. Since X is non-zero, it has some non-vanishing cohomology, say $H^i(X)$. Since kG has a filtration where the filtered quotients are isomorphic to k, $X \otimes_k kG$ is in $\mathsf{Loc}(X)$. Then $H^i(X \otimes_k kG) = H^i(X) \otimes_k kG$ is a free module. So it splits off the complex (Exercise!) and kG concentrated in degree i is in $\mathsf{Loc}(X)$. But this generates $\mathsf{D}(\mathsf{Mod}\,kG)$. $\square$

3.3.4 The category $\mathsf{K}(\mathsf{Inj}\,kG)$

Better than either the stable module category or the derived category is the homotopy category of injective modules $\mathsf{K}(\mathsf{Inj}\,kG)$. The following result is an analogue of Theorem 2.31 for the homotopy category:

Theorem 3.17. *The homotopy category $\mathsf{K}(\mathsf{Inj}\,kG)$ is triangulated, with suspension defined by Σ. Moreover, it is compactly generated, and the natural localisation functor $\mathsf{K}(\mathsf{Inj}\,kG) \to \mathsf{D}(\mathsf{Mod}\,kG)$ induces an equivalence of categories:*

$$\mathsf{K}(\mathsf{Inj}\,kG)^{\mathsf{c}} \simeq \mathsf{D}^{\mathsf{b}}(\mathsf{mod}\,kG)\,.$$

The quasi-inverse associates to each complex in $\mathsf{D}^{\mathsf{b}}(\mathsf{mod}\,kG)$ its injective resolution. In particular, the injective resolutions of the simple modules form a compact generating set for $\mathsf{K}(\mathsf{Inj}\,kG)$.

Proof. The first part, namely that $\mathsf{K}(\mathsf{Inj}\,kG)$ is a triangulated category, is special case of Example 1.42(2). For the statement about compact generation and the identification of compact objects, see [41, Proposition 2.3]. □

In view of this theorem, $\mathsf{K}(\mathsf{Inj}\,kG)$ should be regarded as the correct "big" category for $\mathsf{D}^{\mathsf{b}}(\mathsf{mod}\,kG)$, whereas $\mathsf{D}(\mathsf{Mod}\,kG)$ is not.

Let us give names for some particular objects in $\mathsf{K}(\mathsf{Inj}\,kG)$ that we shall need. We write:

- ik for an injective resolution of k,

- pk for a projective resolution of k, and

- tk for a Tate resolution of k.

In $\mathsf{K}(\mathsf{Inj}\,kG)$ there is then an exact triangle

$$pk \to ik \to tk \to .$$

For the next step, we note that if X and Y are objects in $\mathsf{K}(\mathsf{Inj}\,kG)$, then $X \otimes_k Y$ can be made into an object in $\mathsf{K}(\mathsf{Inj}\,kG)$ by taking the total complex of the tensor product, with diagonal action:

$$(X \otimes_k Y)_i = \bigoplus_{j+k-i} X_j \otimes Y_k$$

with differential

$$d(x \otimes y) = d(x) \otimes y + (-1)^{|x|}x \otimes d(y).$$

The object ik acts as a *tensor identity*:

Proposition 3.18. *For any complex X of injective kG-modules, the map $k \to ik$ induces an isomorphism $X \to X \otimes_k ik$ in $\mathsf{K}(\mathsf{Inj}\,kG)$.* □

3.3.5   Recollement

The relationship between $\mathsf{K}(\mathsf{Inj}\,kG)$, the derived category $\mathsf{D}(\mathsf{Mod}\,kG)$ and the stable module category $\mathsf{StMod}(kG)$ is given by a *recollement*, as follows.

We write $\mathsf{K}_{\mathrm{ac}}(\mathsf{Inj}\,kG)$ for the full subcategory of $\mathsf{K}(\mathsf{Inj}\,kG)$ given by the acyclic complexes of injective kG-modules. Every acyclic complex is a Tate resolution of a module, namely the image of the middle map in the complex. Homotopy classes of maps between acyclic complexes correspond to homomorphisms in the stable module category. Thus Tate resolutions give an equivalence of categories

$$\mathsf{StMod}(kG) \simeq \mathsf{K}_{\mathrm{ac}}(\mathsf{Inj}\,kG)\,.$$

The next result is from [14].

Theorem 3.19. *There is a recollement*

$$\mathsf{K}_{\mathrm{ac}}(\mathsf{Inj}\,kG) \;\underset{-\otimes_k tk}{\overset{\mathrm{Hom}_k(tk,-)}{\rightleftarrows}}\; \mathsf{K}(\mathsf{Inj}\,kG) \;\underset{-\otimes_k pk}{\overset{\mathrm{Hom}_k(pk,-)}{\rightleftarrows}}\; \mathsf{D}(\mathsf{Mod}\,kG). \qquad \square$$

In the diagram, the unlabelled arrow on the left denotes inclusion while the one on the right is the natural quotient functor. Morever, each functor is right adjoint to the one below; thus, for example, $\mathrm{Hom}_k(tk,-)$ is right adjoint to the inclusion functor. Thus the category $\mathsf{K}(\mathsf{Inj}\,kG)$ can be thought of as being glued together from the categories $\mathsf{StMod}(kG)$ and $\mathsf{D}(\mathsf{Mod}\,kG)$.

The compact objects in these categories are only preserved by the left adjoints, since the corresponding right adjoints preserve arbitrary direct sums, giving us the following sequence:

$$\mathsf{stmod}(kG) \xleftarrow{\;-\otimes_k tk\;} \mathsf{D}^{\mathrm{b}}(\mathsf{mod}\,kG) \xleftarrow{\;-\otimes_k pk\;} \mathsf{D}^{\mathrm{b}}(\mathsf{proj}\,kG).$$

This expresses $\mathsf{stmod}(kG)$ as the quotient of $\mathsf{D}^{\mathrm{b}}(\mathsf{mod}\,kG)$ by the perfect complexes. This was first proved by Buchweitz [21]; see also Rickard [52].

3.3.6   Varieties for objects in $\mathsf{K}(\mathsf{Inj}\,kG)$

We now introduce a notion of support for objects in $\mathsf{K}(\mathsf{Inj}\,kG)$. To begin with note that the graded endomorphism ring of ik is $H^*(G,k)$, which is graded commutative and noetherian. For each object X in $\mathsf{K}(\mathsf{Inj}\,kG)$ there is a homomorphism

$$H^*(G,k) = \mathrm{Ext}^*_{kG}(ik,ik) \xrightarrow{-\otimes_k X} \mathrm{Ext}^*_{kG}(X\otimes_k ik, X\otimes_k ik) \cong \mathrm{Ext}^*_{kG}(X,X)\,,$$

where the isomorphism is from Proposition 3.18. This allows us to apply the technology developed in Sections 3.1 and 3.2 in this context.

Let $\mathcal{V}_G$ be the set of homogeneous prime ideals in $H^*(G,k)$ (including the maximal one). Then for each $\mathfrak{p} \in \mathcal{V}_G$ we have a local cohomology functor

$$\Gamma_{\mathfrak{p}} \colon \mathsf{K}(\mathsf{Inj}\,kG) \to \mathsf{K}(\mathsf{Inj}\,kG)\,.$$

Definition 3.20. For an object X in $\mathsf{K}(\mathsf{Inj}\,kG)$, we define

$$\mathcal{V}_G(X) = \{\mathfrak{p} \in \mathcal{V}_G \mid \Gamma_\mathfrak{p} X \neq 0\}.$$

The assertions in the result below are all obtained as special cases of results concerning local cohomology and support for triangulated categories, discussed in the previous lectures.

Proposition 3.21. *The assignment $X \mapsto \mathcal{V}_G(X)$ has the following properties:*

(1) $\mathcal{V}_G(X) = \varnothing$ *if and only if $X = 0$.*

(2) $\mathcal{V}_G(X \oplus Y) = \mathcal{V}_G(X) \cup \mathcal{V}_G(Y)$, *and more generally*

$$\mathcal{V}_G\Big(\bigoplus_\alpha X_\alpha\Big) = \bigcup_\alpha \mathcal{V}_G(X_\alpha).$$

(3) $\mathcal{V}_G(\Gamma_\mathfrak{p} ik) = \{\mathfrak{p}\}$.

(4) *For any $\mathcal{V} \subseteq \mathcal{V}_G$, there exists an object X in $\mathsf{K}(\mathsf{Inj}\,kG)$ with $\mathcal{V}_G(X) = \mathcal{V}$.* $\square$

Properties of support that are specific to $\mathsf{K}(\mathsf{Inj}\,kG)$, including a tensor product formula for supports:

$$\mathcal{V}_G(X \otimes_k Y) = \mathcal{V}_G(X) \cap \mathcal{V}_G(Y)$$

will be deduced as a consequence of results presented in later lectures.

3.3.7 Comparison with $\mathsf{StMod}(kG)$

For objects in the subcategory

$$\mathsf{K}_{\mathrm{ac}}(\mathsf{Inj}\,kG) \simeq \mathsf{StMod}(kG)$$

of $\mathsf{K}(\mathsf{Inj}\,kG)$ the definition of $\mathcal{V}_G(X)$ given in the previous section coincides with that of Definition 2.35. Since the object k in $\mathsf{StMod}(kG)$ corresponds to tk in $\mathsf{K}_{\mathrm{ac}}(\mathsf{Inj}\,kG)$ we write $\mathcal{V}_G(tk)$ for what was denoted $\mathcal{V}_G(k)$ in the discussion of varieties for $\mathsf{StMod}(kG)$. Thus we have

$$\mathcal{V}_G = \mathcal{V}_G(tk) \cup \{H^{\geqslant 1}(G, k)\}$$

where $H^{\geqslant 1}(G, k)$ is the maximal ideal of positive degree elements.

Referring back to the recollement (Theorem 3.19), X is isomorphic to $X \otimes_k tk$ if and only if X is in $\mathsf{K}_{\mathrm{ac}}(\mathsf{Inj}\,kG)$. If X is not acyclic, then

$$\mathcal{V}_G(X) = \mathcal{V}_G(X \otimes_k tk) \cup \{H^{\geqslant 1}(G, k)\}\,.$$

Remark 3.22. It is not necessary to understand the rest of this lecture for the goals of this seminar. Our purpose is to place $\mathsf{K}(\mathsf{Inj}\,kG)$ in a wider context.

60Chapter 3. Wednesday

3.3.8 $\mathsf{K}(\operatorname{Inj} B)$ as a derived invariant

Recall that the *blocks* of a group algebra kG are the indecomposable two sided
ideal direct factors. So the block decomposition of kG is of the form

$$B_0 \times \cdots \times B_s\,.$$

This decomposition is unique, and every indecomposable kG-module is a module
for B_i for a unique value of i.

Theorem 3.23. *Let B and B' be blocks of group algebras. The following conditions
are equivalent:*

(1) *There is a tilting complex over B whose endomorphism ring in $\mathsf{D}^{\mathsf{b}}(\operatorname{mod} B)$ is
    isomorphic to B'*

(2) $\mathsf{D}^{\mathsf{b}}(\operatorname{mod} B)$ *and* $\mathsf{D}^{\mathsf{b}}(\operatorname{mod} B')$ *are triangle equivalent*

(3) $\mathsf{D}(\operatorname{Mod} B)$ *and* $\mathsf{D}(\operatorname{Mod} B')$ *are triangle equivalent*

(4) $\mathsf{K}(\operatorname{Inj} B)$ *and* $\mathsf{K}(\operatorname{Inj} B')$ *are triangle equivalent.* $\qquad\square$

The equivalence of the first three of these is Rickard's theorem.

3.3.9 $\mathsf{K}(\operatorname{Inj} kG)$ is a derived category

Given complexes X and Y of kG-modules, we form a complex $\mathcal{H}om_{kG}(X,Y)$ whose
nth component is

$$\prod_{m\in\mathbb{Z}} \mathcal{H}om_{kG}(X_m, Y_{n+m})$$

with differential given by

$$(d(f))(x) = d(f(x)) - (-1)^{|f|} f(d(x))\,.$$

Composition of maps makes $\mathcal{E}nd_{kG}(X) = \mathcal{H}om_{kG}(X,X)$ into a *differential graded
algebra* over which $\mathcal{H}om_{kG}(X,Y)$ is a *differential graded module*.

Theorem 3.24. *If C is a compact generator for $\mathsf{K}(\operatorname{Inj} kG)$, then*

$$\mathcal{H}om_{kG}(C,-)\colon \mathsf{K}(\operatorname{Inj} kG) \to \mathsf{D}(\mathcal{E}nd_{kG}(C))$$

is an equivalence of categories.

If C is just a compact object, which does not necessarily generate, then

$$\mathcal{H}om_{kG}(C,-)\colon \operatorname{Loc}(C) \to \mathsf{D}(\mathcal{E}nd_{kG}(C))$$

is an equivalence of categories.

Proof. This is a direct application of Monday's Exercise 23. $\qquad\square$

In this theorem, we have used $\mathsf{D}(A)$ to denote the *derived category* of a differential graded algebra A. To form this, we first form the category whose objects are the differential graded modules and whose arrows are the homotopy classes of degree preserving maps. Then we invert the quasi-isomorphisms just as we did in the usual derived category of a ring. For more details, see Section 5.2.1.

If the differential graded algebra is just a ring concentrated in degree zero, with zero differential, then a differential graded module is the same as a complex of modules and we recover the usual definition of the derived category of a ring.

Observation 3.25. If G is a p-group, then k is the only simple kG-module, so that ik is a compact generator for $\mathsf{K}(\mathsf{Inj}\, kG)$; see Theorem 3.17. Therefore

$$\mathsf{K}(\mathsf{Inj}\, kG) \simeq \mathsf{D}(\mathcal{E}nd_{kG}(ik)) \,.$$

For a non-p-group, we just get an equivalence between the localising subcategory of $\mathsf{K}(\mathsf{Inj}\, kG)$ generated by ik and $\mathsf{D}(\mathcal{E}nd_{kG}(ik))$.

The next result, which goes by the name *Rothenberg–Steenrod construction*, provides a link to algebraic topology:

Theorem 3.26. *For any path-connected space X we have a quasi-isomorphism of differential graded algebras*

$$\mathcal{E}nd_{C_*(\Omega X;k)}(k) \simeq C^*(X;k)$$

where ΩX is the loop space of X. □

In case $X = BG$, the connected components of ΩX are contractible, and we have $\Omega X \simeq G$. So the differential graded algebra $C_*(\Omega X;k)$ is quasi-isomorphic to the group algebra kG. So the Rothenberg–Steenrod construction gives

$$\mathcal{E}nd_{C_*(\Omega X;k)}(k) \simeq \mathcal{E}nd_{kG}(ik) \simeq C^*(BG;k).$$

Theorem 3.27. *If G is a finite p-group, there are equivalences of categories*

$$\mathsf{K}(\mathsf{Inj}\, kG) \simeq \mathsf{D}(\mathcal{E}nd_{kG}(ik)) \simeq \mathsf{D}(C^*(BG;k)) \,.$$

For a more general finite group, the same argument shows that the right side is equivalent to the subcategory $\mathrm{Loc}(ik)$ of $\mathsf{K}(\mathsf{Inj}\, kG)$.

Under this equivalence the tensor product $- \otimes_k -$ with diagonal G-action on the left-hand side corresponds to the derived E_∞ tensor product on the right-hand side. This is remarkable, for the latter takes a great deal of topological machinery to develop, see for example Elmendorf, Kriz, Mandell, and May [28].

We have the following dictionary relating operations in $\mathsf{K}(\mathsf{Inj}\, kG)$ with operations in $\mathsf{D}(C^*(BG;k))$. Note that this dictionary turns things upside down, so that induction corresponds to restriction and restriction corresponds to coinduction.

K(Inj kG)	D($C^*(BG;k)$)
$- \otimes_k -$ diagonal G-action	$- \otimes^{\mathbf{L}}_{C^*(BG;k)} -$ E_∞ tensor product
induction, $-_H{\uparrow}^G$ $kG \otimes_{kH} -$	restriction via $C^*(BG;k) \to C^*(BH;k)$
restriction, $-_G{\downarrow}_H$ via $kH \to kG$	coinduction $\mathrm{Hom}_{C^*(BG;k)}(C^*(BH;k),-)$
ik	$C^*(BG;k)$
kG	k

3.4   Exercises

Take a hike and eat a cake.

4 Thursday

In the last chapter we introduced a notion of support for triangulated categories. The starting point in this one is a notion that we call 'stratification' for an R-linear triangulated category T. It identifies conditions under which support can be used to parameterise localising subcategories of T. The crux of the stratification condition is local in nature, in that it involves only the subcategories $\Gamma_{\mathfrak{p}}\mathsf{T}$, so can be, and usually is, verified one prime at a time. We illustrate this technique by outlining the proof of the main results of this seminar; all this is part of Section 4.1. At first glance, the stratification condition is rather technical and of limited scope. To counter this, in Section 4.2 we discuss a number of interesting consequences that follow from this property. The last section has a different flavour: it makes concrete some of the ideas and constructions we have been discussing by describing them in the case of the Klein four group.

4.1 Stratifying triangulated categories

In this lecture we identify under what conditions the notion of support classifies localising subcategories of a triangulated category. Most of the material in this lecture is based on [12].

4.1.1 Classifying localising subcategories

Recall the setting from Wednesday's lectures: T is an R-linear triangulated category, meaning that T is a compactly generated triangulated category with small coproducts and R is a noetherian graded commutative ring with a given homomorphism $R \to Z^*(\mathsf{T})$ to the graded centre of T. This amounts to giving for each object X in T a homomorphism of graded rings

$$R \to \mathrm{End}_{\mathsf{T}}^*(X),$$

and these homomorphisms satisfy the following compatibility condition. Given a morphism $\phi\colon X \to Y$ in T, the diagram

$$
\begin{array}{ccc}
R & \longrightarrow & \operatorname{End}^*_{\mathsf{T}}(X) \\
\big\downarrow & & \big\downarrow {\scriptstyle \operatorname{Hom}^*_{\mathsf{T}}(X,\phi)} \\
\operatorname{End}^*_{\mathsf{T}}(Y) & \xrightarrow{\ \operatorname{Hom}^*_{\mathsf{T}}(\phi,Y)\ } & \operatorname{Hom}^*_{\mathsf{T}}(X,Y)
\end{array}
$$

commutes, up to the expected sign. We set

$$\operatorname{Spec} R = \{\text{homogeneous prime ideals of } R\}.$$

For each $\mathfrak{p} \in \operatorname{Spec} R$ there is a *local cohomology functor* which is an exact functor $\Gamma_{\mathfrak{p}}\colon \mathsf{T} \to \mathsf{T}$. For each object X in T there is a *support*

$$\operatorname{supp}_R X = \{\mathfrak{p} \in \operatorname{Spec} R \mid \Gamma_{\mathfrak{p}} X \neq 0\}.$$

The support of T is the set

$$\operatorname{supp}_R \mathsf{T} = \{\mathfrak{p} \in \operatorname{Spec} R \mid \Gamma_{\mathfrak{p}} \mathsf{T} \neq 0\}.$$

Our goal in this lecture is to examine under what conditions this notion of support classifies localising subcategories. We have a map

$$\{\text{localising subcategories of } \mathsf{T}\} \xrightarrow{\sigma} \{\text{subsets of } \operatorname{supp}_R T\}$$

given by

$$\mathsf{S} \mapsto \bigcup_{X \in \mathsf{S}} \operatorname{supp}_R X.$$

When this map is a bijection, we say that support *classifies* localising subcategories.

Remark 4.1. If σ is a bijection, then the inverse map

$$\{\text{localising subcategories of } \mathsf{T}\} \xleftarrow{\tau} \{\text{subsets of } \operatorname{supp}_R T\}$$

sends a subset $\mathcal{V}$ of $\operatorname{supp}_R \mathsf{T}$ to the localising subcategory consisting of the modules whose support is contained in $\mathcal{V}$.

Proposition 4.2. *If σ is a bijection, there are two consequences:*

(1) *The* local-global principle: *for each object X, the localising subcategory generated by X is the same as that generated by*

$$\{\Gamma_{\mathfrak{p}} X \mid \mathfrak{p} \in \operatorname{Spec} R\}.$$

(2) *The* minimality condition: *for $\mathfrak{p} \in \operatorname{supp}_R \mathsf{T}$, the subcategory $\Gamma_{\mathfrak{p}} \mathsf{T}$ of objects supported at $\mathfrak{p}$ is a* minimal *localising subcategory of T.*

Definition 4.3. We say that T is *stratified* by the action of R if the two conditions listed in Proposition 4.2 hold.

We should note right away that the minimality is the critical condition; we know of no examples where the local-global principle fails; see Section 4.1.3.

One of the crucial observations of [12] is the converse to Proposition 4.2: if these two conditions hold, then the map is a bijection. The map $\sigma\tau$ is clearly the identity. To see that $\tau\sigma$ is the identity, we use both the local-global principle and the minimality condition. We will prove a more general statement in Theorem 4.5, later in this lecture.

4.1.2 Minimality

It is useful to have a criterion for checking the minimality condition. The following test is Lemma 4.1 of [12].

Lemma 4.4. *Assume that* T *is compactly generated. A non-zero localising subcategory* S *of* T *is minimal if and only if for every pair of non-zero objects* X *and* Y *in* S *we have* $\mathrm{Hom}_{\mathsf{T}}^*(X, Y) \neq 0$.

Proof. If S is minimal, then $Y \in \mathrm{Loc}_{\mathsf{T}}(X)$ so if $\mathrm{Hom}_{\mathsf{T}}^*(X, Y) = 0$ it would follow that $\mathrm{Hom}_{\mathsf{T}}^*(Y, Y) = 0$ and so $Y = 0$.

Conversely if S has a proper non-zero localising subcategory S′, we may assume $S' = \mathrm{Loc}_{\mathsf{T}}(X)$ for some $X \neq 0$. Since T is compactly generated, there is a localisation functor for S′ (Lemma 2.1 of [12]). So if $W \in \mathsf{S} \setminus \mathsf{S}'$, we have a triangle $W' \to W \to W'' \to$ with $W' \in \mathsf{S}'$ (so $W'' \neq 0$) and $\mathrm{Hom}_{\mathsf{T}}(X, W'') = 0$. □

What happens if just the local-global principle holds, without assuming that the minimality condition holds? In this case, we have to take into account the localising subcategories of $\Gamma_{\mathfrak{p}}\mathsf{T}$.

Theorem 4.5. *Set* $\mathcal{V} = \mathrm{supp}_R \mathsf{T}$. *When the local-global principle holds, there are inclusion preserving bijections*

$$\left\{ \begin{array}{c} \text{\textit{Localizing}} \\ \text{\textit{subcategories of}} \ \mathsf{T} \end{array} \right\} \underset{\tau}{\overset{\sigma}{\underset{\longleftarrow}{\longrightarrow}}} \left\{ \begin{array}{c} \text{\textit{Families}} \ (\mathsf{S}(\mathfrak{p}))_{\mathfrak{p}\in\mathcal{V}} \ \text{\textit{with}} \ \mathsf{S}(\mathfrak{p}) \ a \\ \text{\textit{localizing subcategory of}} \ \Gamma_{\mathfrak{p}}\mathsf{T} \end{array} \right\}$$

defined by $\sigma(\mathsf{S}) = (\mathsf{S} \cap \Gamma_{\mathfrak{p}}\mathsf{T})_{\mathfrak{p}\in\mathcal{V}}$ *and* $\tau(\mathsf{S}(\mathfrak{p}))_{\mathfrak{p}\in\mathcal{V}} = \mathrm{Loc}_{\mathsf{T}}(\mathsf{S}(\mathfrak{p}) \mid \mathfrak{p} \in \mathcal{V})$.

Proof. We first claim that $\sigma\tau$ is the identity. Set $\mathsf{S} = \mathrm{Loc}(\mathsf{S}(\mathfrak{p}) \mid \mathfrak{p} \in \mathcal{V})$. Since $\Gamma_{\mathfrak{p}}\Gamma_{\mathfrak{q}} = 0$ for $\mathfrak{q} \neq \mathfrak{p}$, we have $\Gamma_{\mathfrak{p}}\mathsf{S} = \mathsf{S}(\mathfrak{p})$. Since $\Gamma_{\mathfrak{p}}\Gamma_{\mathfrak{p}} = \Gamma_{\mathfrak{p}}$ we therefore have $\mathsf{S} \cap \Gamma_{\mathfrak{p}}\mathsf{T} \subseteq \Gamma_{\mathfrak{p}}\mathsf{S} = \mathsf{S}(\mathfrak{p})$. The reverse inclusion is obvious.

Next we claim that $\tau\sigma$ is the identity. Let S be a localising subcategory of T. We have $\tau\sigma(\mathsf{S}) = \mathrm{Loc}_{\mathsf{T}}(\mathsf{S} \cap \Gamma_{\mathfrak{p}}\mathsf{T} \mid \mathfrak{p} \in \mathcal{V}) \subseteq \mathsf{S}$. So letting X be in S, we must prove that X is in $\tau\sigma(\mathsf{S})$. Using the local-global principle, we have

$$\Gamma_{\mathfrak{p}}X \in \mathsf{S} \cap \Gamma_{\mathfrak{p}}\mathsf{T} \subseteq \tau\sigma(\mathsf{S}),$$

and so $X \in \mathrm{Loc}_{\mathsf{T}}(\Gamma_{\mathfrak{p}}X \mid \mathfrak{p} \in \mathrm{supp}_R X) \subseteq \tau\sigma(\mathsf{S})$. □

4.1.3   When does the local-global principle hold?

Turning now to the local-global principle, one has the following:

Theorem 4.6. *Let* T *be compactly generated and* X *an object in* T *satisfying* $\dim \operatorname{supp}_R X < \infty$. *Then* $\operatorname{Loc}_{\mathsf{T}}(X) = \operatorname{Loc}_{\mathsf{T}}\{\Gamma_{\mathfrak{p}} X \mid \mathfrak{p} \in \operatorname{supp}_R X\}$. $\qquad\square$

As an immediate consequence, one gets:

Corollary 4.7. *If* $\dim \operatorname{Spec} R$ *is finite, then the local-global principle holds.* $\qquad\square$

The local-global principle also holds automatically in the context of tensor triangulated categories. We discuss these next.

4.1.4   Tensor triangulated categories

Let $(\mathsf{T}, \otimes, \mathbb{1})$ be a *tensor triangulated category*. For us, this means:

- T is a triangulated category,

- $\otimes\colon \mathsf{T} \times \mathsf{T} \to \mathsf{T}$ is a *symmetric monoidal* tensor product (i.e., commutative and associative up to coherent natural isomorphisms),

- $\otimes$ is exact in each variable and preserves small coproducts,

- $\mathbb{1}$ is a unit for the tensor product, and is compact.

The *Brown representability theorem* implies that there exist *function objects*. Given objects X and Y in T there is an object $\mathcal{H}om(X, Y)$ in T, contravariant in X and covariant in Y, together with an adjunction

$$\operatorname{Hom}_{\mathsf{T}}(X \otimes Y, Z) \cong \operatorname{Hom}_{\mathsf{T}}(X, \mathcal{H}om(Y, Z)).$$

By construction, $\mathcal{H}om(X, Y)$ will be functorial in X, and we assume also functoriality in Y; all examples encountered in these notes have this property.

Since T is tensor triangulated category, the ring $\operatorname{End}^*_{\mathsf{T}}(\mathbb{1})$ is *graded commutative*. It acts on T via

$$\operatorname{End}^*_{\mathsf{T}}(\mathbb{1}) \xrightarrow{\ -\otimes X\ } \operatorname{End}^*_{\mathsf{T}}(X).$$

If R is noetherian and graded commutative, then any homomorphism $R \to \operatorname{End}^*_{\mathsf{T}}(\mathbb{1})$ of rings induces an action $R \to Z(\mathsf{T})$. Such an action is said to be *canonical*.

If the action of R is canonical, the adjunction defining function objects is R-linear. Using this, we get

$$\Gamma_{\mathcal{V}} X \cong X \otimes \Gamma_{\mathcal{V}} \mathbb{1}, \qquad L_{\mathcal{V}} X \cong X \otimes L_{\mathcal{V}} \mathbb{1}, \qquad \Gamma_{\mathfrak{p}} X \cong X \otimes \Gamma_{\mathfrak{p}} \mathbb{1}.$$

Here, $\mathcal{V}$ is a specialisation closed subset and $\mathfrak{p}$ is a point in $\operatorname{Spec} R$.

In a tensor triangulated category, we talk of *tensor ideal* localising subcategories: S is tensor ideal if $X \in \mathsf{S}$, $Y \in \mathsf{T}$ implies $X \otimes Y \in \mathsf{S}$.

Definition 4.8. We write $\mathrm{Loc}_{\mathsf{T}}^{\otimes}(\mathsf{S})$ for the smallest tensor ideal localising subcategory of T containing a subcategory (or a collection of objects) S.

The following result is [12, Theorem 7.2].

Theorem 4.9. *Let* T *be a tensor triangulated category with canonical R-action. Then for each object X in T we have*

$$\mathrm{Loc}_{\mathsf{T}}^{\otimes}(X) = \mathrm{Loc}_{\mathsf{T}}^{\otimes}(\Gamma_{\mathfrak{p}}X \mid \mathfrak{p} \in \mathrm{Spec}\, R).$$

In particular, if $\mathbb{1}$ generates T, then the local global principle holds for T. $\square$

In a tensor triangulated category T with a canonical action of a graded commutative noetherian ring R, the localising subcategories defined by the functors $\Gamma_{\mathfrak{p}}$ are all tensor ideal. It therefore makes sense to restrict our attention to the classification of tensor ideal localising subcategories. In this context, we have the following analogue of Theorem 4.5.

Theorem 4.10. *Let* T *be a tensor triangulated category with canonical R-action. Set $\mathcal{V} = \mathrm{supp}_R \mathsf{T}$. Then there is bijection*

$$\left\{ \begin{array}{c} \textit{Tensor ideal localizing} \\ \textit{subcategories of } \mathsf{T} \end{array} \right\} \underset{\tau}{\overset{\sigma}{\rightleftarrows}} \left\{ \begin{array}{c} \textit{Families } (\mathsf{S}(\mathfrak{p}))_{\mathfrak{p}\in\mathcal{V}} \textit{ with } \mathsf{S}(\mathfrak{p}) \textit{ a tensor} \\ \textit{ideal localizing subcategory of } \Gamma_{\mathfrak{p}}\mathsf{T} \end{array} \right\}.$$

The map σ and its inverse τ are defined in the same way as in Theorem 4.5, and the proof is essentially the same as the proof of that theorem.

Since the local-global principle automatically holds for tensor ideal localising subcategories, we say that a tensor triangulated category T is *stratified* by R if every $\Gamma_{\mathfrak{p}}\mathsf{T}$ is either minimal among tensor ideal localising subcategories or is zero. Under this condition, the maps σ and τ establish a bijection between tensor ideal localising subcategories of T and subsets of $\mathrm{supp}_R \mathsf{T}$.

The analogue of the minimality test of Lemma 4.4 in the tensor triangulated situation is as follows.

Lemma 4.11. *Assume that T is a compactly generated tensor triangulated category with small coproducts. A non-zero tensor ideal localising subcategory S of T is minimal if and only if for every pair of non-zero objects X and Y in S there exists an object Z such that we have $\mathrm{Hom}_{\mathsf{T}}^{*}(X \otimes Z, Y) \neq 0$.* $\square$

In fact, the object Z can be chosen independently of X and Y. For example, a compact generator for T always suffices.

4.1.5 The stable module category

The category $\mathsf{StMod}(kG)$ is a tensor triangulated category, so the local-global principal holds. In [11] we prove:

Theorem 4.12. *The tensor triangulated category $\mathsf{StMod}(kG)$ is stratified by the canonical action of $H^{*}(G, k)$.* $\square$

What we have seen in this lecture is that in order to prove this, we only need to see that each $\Gamma_{\mathfrak{p}}\,\mathsf{StMod}(kG)$ is minimal or zero. In fact the only case in which we get zero is if $\mathfrak{p}$ is the maximal ideal of positive degree elements. So the support of $\mathsf{StMod}(kG)$ is the set of non-maximal homogeneous primes in $\mathcal{V}_G = \operatorname{Spec} H^*(G, k)$. As a direct consequence one gets:

Corollary 4.13. *Support defines a one-to-one correspondence between tensor ideal localising subcategories of* $\mathsf{StMod}(kG)$ *and subsets of the set of non-closed points in* $\mathcal{V}_G$. $\qquad\square$

We prove Theorem 4.12 by first establishing the corresponding result for the homotopy category, $\mathsf{K}(\mathsf{Inj}\,kG)$, discussed next.

4.1.6   The category $\mathsf{K}(\mathsf{Inj}\,kG)$

Again this is a tensor triangulated category, so the local-global principal holds. The following is the main result of [11]:

Theorem 4.14. *The tensor triangulated category* $\mathsf{K}(\mathsf{Inj}\,kG)$ *is stratified by the canonical action of* $H^*(G, k)$. $\qquad\square$

This time, every point in $\mathcal{V}_G$, the homogenous spectrum of $H^*(G, k)$, including the maximal ideal, appears in the support of $\mathsf{K}(\mathsf{Inj}\,kG)$.

Corollary 4.15. *Support defines a one-to-one correspondence between tensor ideal localising subcategories of* $\mathsf{K}(\mathsf{Inj}\,kG)$ *and subsets of the set of all subsets of* $\mathcal{V}_G$. $\quad\square$

4.1.7   Overview of classification for $\mathsf{Mod}(kG)$

We finish this lecture with an outline of the strategy for proving the classification theorem for subcategories of $\mathsf{Mod}(kG)$ satisfying the two conditions of Definition 1.18 together with the tensor ideal condition. The main theorem states that these are in one-to-one correspondence with subsets of the set of non-maximal homogeneous prime ideals in $H^*(G, k)$.

Step 1. Go down from $\mathsf{Mod}(kG)$ to $\mathsf{StMod}(kG)$. Every non-zero subcategory of $\mathsf{Mod}(kG)$ satisfying the given conditions contains the projective modules, and passes down to a tensor ideal localising subcategory of $\mathsf{StMod}(kG)$. This gives a bijection with the set of tensor ideal localising subcategories of $\mathsf{StMod}(kG)$, so we are reduced to proving Theorem 4.12.

Step 2. Go up to $\mathsf{K}(\mathsf{Inj}\,kG)$. According to the discussion in Lecture 3, each tensor ideal localising subcategory of $\mathsf{StMod}(kG)$ corresponds to two such for $\mathsf{K}(\mathsf{Inj}\,kG)$. One is the image under the inclusion

$$\mathsf{StMod}(kG) \simeq \mathsf{K}_{\mathrm{ac}}(\mathsf{Inj}\,kG) \to \mathsf{K}(\mathsf{Inj}\,kG)$$

and the other is generated by this together with the extra object pk. So we are reduced to proving Theorem 4.14.

Step 3. Reduce from $\mathsf{K}(\mathsf{Inj}\,kG)$ to $\mathsf{K}(\mathsf{Inj}\,kE)$ for E an elementary abelian p-group using the Quillen Stratification Theorem and a suitably strengthened version of Chouinard's Theorem. This is the subject of the last lecture on Friday.

Step 4. If $p = 2$, we can go from $\mathsf{K}(\mathsf{Inj}\,kE)$ to differential graded modules over a graded polynomial ring, viewed as a differential graded algebra, using a version of the Bernstein–Gelfand–Gelfand correspondence. This is explained in Section 5.2.

The proof is rather more complicated when p is odd. We still get down to a graded polynomial ring, but the reduction involves a number of steps. The first one is to pass to the Koszul complex of the group algebra kE. The Koszul complex can be viewed as a differential graded algebra, and then it is quasi-isomorphic to a graded exterior algebra, since kE is a complete intersection ring. We then again apply a version of the Bernstein–Gelfand–Gelfand correspondence to get to a graded polynomial ring. The technical tools needed to execute this proof require considerable preparation, beyond what is already presented in these lectures, so we can do no more than refer the interested readers to the source [11].

Step 5. Deal directly with a graded polynomial ring using the minimality condition discussed in this lecture. This is addressed in Section 5.1.

4.2 Consequences of stratification

In the last lecture we learnt about the stratification condition and its connection to the problem of classifying localizing subcategories of triangulated categories. In this lecture we present some other, not immediately obvious, consequences of stratification that serve to illustrate how strong a condition it is, and how useful it is when one can establish that it holds in some context. The basic reference for the material presented below is again [12].

For most of the lecture T will be an R-linear triangulated category as in Section 3.1; we tackle the tensor triangulated case at the very end. Recall that T is said to be stratified by R if:

(S1) The local global principle holds; see 4.2. An equivalent condition is that

$$\Gamma_{\mathcal{V}} X \in \mathsf{Loc}_{\mathsf{T}}(\Gamma_{\mathfrak{p}} X \mid \mathfrak{p} \in \mathrm{supp}_R X)$$

for any specialisation closed subset $\mathcal{V}$ of $\mathrm{Spec}\,R$; see [12, Theorem 3.1].

(S2) $\Gamma_{\mathfrak{p}} \mathsf{T}$ is minimal for each $\mathfrak{p} \in \mathrm{supp}_R \mathsf{T}$.

Recall that an object X of T is in $\Gamma_{\mathfrak{p}} \mathsf{T}$ if and only if $H_C^*(X)$ is $\mathfrak{p}$-local and $\mathfrak{p}$-torsion for all $C \in \mathsf{T}^c$; see Corollary 3.9. Condition (S2) is the statement that for any such $X \neq 0$, one has $\mathsf{Loc}_{\mathsf{T}}(X) = \Gamma_{\mathfrak{p}} \mathsf{T}$; that is to say, X builds any other $\mathfrak{p}$-local and $\mathfrak{p}$-torsion object.

Remark 4.16. The local global principle (S1) holds when the Krull dimension of R is finite, or if T is tensor triangulated with $\mathrm{Loc}_\mathsf{T}(\mathbb{1}) = \mathsf{T}$ and the ring R acts on T via a homomorphism $R \to \mathrm{End}^*_\mathsf{T}(\mathbb{1})$.

Example 4.17. When A is a commutative noetherian ring, $\mathsf{D}(A)$ is stratified by the canonical A-action. Indeed, (S1) holds because $\mathsf{D}(A)$ is suitably tensor triangulated, while (S2) is by a theorem of Neeman [44]. We will present a proof of this result in Section 5.1.

Example 4.18. When G is a finite group and k a field with char k dividing $|G|$, both $\mathsf{StMod}(kG)$ and $\mathsf{K}(\mathsf{Inj}\,kG)$ are stratified by canonical actions of the cohomology ring $H^*(G;k)$. This is proved in [11], and is the focal point of this seminar.

4.2.1   Classification theorems

For the remainder of this lecture T will be a triangulated category stratified by an action of a graded commutative ring R. The first consequence is that localising subcategories are parameterised by subsets of $\mathrm{supp}_R \mathsf{T}$, which is something that was discussed already in the previous lecture.

Theorem 4.19. *The maps assigning a subcategory* S *to its support,* $\mathrm{supp}_R \mathsf{S}$, *induces a bijection*

$$\{Localizing\ subcategories\ of\ \mathsf{T}\} \xrightarrow{\;\mathrm{supp}_R(-)\;} \{Subsets\ of\ \mathrm{supp}_R \mathsf{T}\}\ .$$

Its inverse sends a subset $\mathcal{U}$ *of* $\mathrm{supp}_R \mathsf{T}$ *to* $\{X \in \mathsf{T} \mid \mathrm{supp}_R X \subseteq \mathcal{U}\}$. $\qquad\square$

The following statement is an immediate consequence of the theorem; on the other hand, it is not hard to prove that the corollary implies stratification.

Corollary 4.20. *If* $\mathrm{supp}_R X \subseteq \mathrm{supp}_R Y$, *then* X *is in* $\mathrm{Loc}_\mathsf{T}(Y)$. $\qquad\square$

Next we explain how, under further, though mild, hypotheses on T stratification implies a classification of the thick subcategories of compact objects.

Definition 4.21. We say that an R-linear triangulated category T is *noetherian* if for all compact objects C in T the R-module $\mathrm{End}^*_\mathsf{T}(C)$ is finitely generated; equivalently, if the R-module $\mathrm{Hom}^*_\mathsf{T}(C,D)$ is finitely generated for all D compact.

The derived category of a commutative noetherian ring R, viewed as an R-linear category, is noetherian: This is easy to verify, once one accepts that the compact objects are the perfect complexes; see Theorem 2.2. Another example of a noetherian category is $\mathsf{K}(\mathsf{Inj}\,kG)$, the homotopy category of complexes of injectives of a finite group G, viewed as an $H^*(G,k)$-linear category. This is a restatement of Even's theorem 2.17, given the identification of compact objects in the categories, Theorem 3.17. Note that the stable module category $\mathsf{StMod}(kG)$ is usually not noetherian, see Theorem 3.14; however, it embeds in the noetherian category $\mathsf{K}(\mathsf{Inj}\,kG)$, by Theorem 3.19, and this usually suffices for applications.

One consequence of the noetherian condition is that supports of compact objects are closed. A more precise statement is proved in [10, Theorem 5.5].

Theorem 4.22. *If C is a compact object in T and the R-module $\mathrm{End}_\mathsf{T}^*(C)$ is noetherian, then*
$$\mathrm{supp}_R C = \mathcal{V}(\mathfrak{a}) \quad \text{where } \mathfrak{a} = \mathrm{Ker}(R \to \mathrm{End}_\mathsf{T}^*(C)).$$
In particular, $\mathrm{supp}_R C$ is a closed subset of $\mathrm{Spec}\,R$. □

From the stratification condition on T one gets:

Theorem 4.23. *When the R-linear triangulated category T is stratified and noetherian, there are inclusion preserving bijections*

$$\left\{ \begin{array}{c} \textit{Thick subcategories} \\ \textit{of compact objects in } \mathsf{T} \end{array} \right\} \begin{array}{c} \xrightarrow{\ \mathrm{supp}_R(-)\ } \\ \xleftarrow[\mathrm{supp}_R^{-1}(-)]{} \end{array} \left\{ \begin{array}{c} \textit{Specialisation closed} \\ \textit{subsets of } \mathrm{supp}_R\,\mathsf{T} \end{array} \right\}$$

where $\mathrm{supp}_R^{-1}(\mathcal{V}) = \{C \in \mathsf{T}^c \mid \mathrm{supp}_R C \subseteq \mathcal{V}\}$, for $\mathcal{V}$ specialisation closed.

Proof. There are two crucial issues: One is that for compact objects C, D, if $\mathrm{supp}_R C \subseteq \mathrm{supp}_R D$, then C is in $\mathrm{Thick}_\mathsf{T}(D)$; second, any closed subset of $\mathrm{supp}_R\,\mathsf{T}$ is the support of some compact object. We leave the latter to the exercises, and sketch an argument for the former.

When $\mathrm{supp}_R C \subseteq \mathrm{supp}_R D$ holds, it follows from Corollary 4.20 that C is in the localising subcategory generated by D. Since C, D are compact, Theorem 1.49 implies that C is in fact in the thick subcategory generated by D, as desired.

A different proof for this part of the proof is presented in [12, Section 6]. □

4.2.2 Orthogonality

A central problem in any additive category is to understand when there are non-zero morphisms between a given pair of objects. Lemma 4.4 makes it clear that this is at the heart of the stratification property for the triangulated categories we have been considering. Conversely, the stratification condition allows us to give fairly precise answers to this problem. The one below is from [12, Section 5]. For a proof in the case of tensor triangulated categories, see Theorem 4.26 and the remarks following it.

Theorem 4.24. *When the R-linear triangulated category T is noetherian and stratified, there is an equality*
$$\mathrm{supp}_R \mathrm{Hom}_\mathsf{T}^*(C, D) = \mathrm{supp}_R C \cap \mathrm{supp}_R D,$$
for each pair of compact objects C, D in T. □

The support on the left-hand side is the usual one from commutative algebra, as the R-module $\mathrm{Hom}_\mathsf{T}^*(C, D)$ is finitely generated; see Lemma A.15.

As a corollary one gets the following "symmetry of Ext vanishing" type result that was proved for local complete intersection rings by Avramov and Buchweitz [1]. The corresponding result is also true for modules over group algebras, and is much easier to prove; see Exercise 10 at the end of this chapter.

Corollary 4.25. *When in addition $R^i = 0$ holds for $i < 0$, one has $\mathrm{Hom}_{\mathsf{T}}^n(C, D) = 0$ for $n \gg 0$ if and only if $\mathrm{Hom}_{\mathsf{T}}^n(D, C) = 0$ for $n \gg 0$.*

Proof. For R as in the statement, a finitely generated R-module M satisfies $M^i = 0$ for $i \gg 0$ if and only if $\mathrm{supp}_R M \subseteq \mathcal{V}(R^{\geqslant 1})$; we leave this as an exercise.

Assume $\mathrm{Hom}_{\mathsf{T}}^n(C, D) = 0$ for $n \geqslant s$. Then $R^{\geqslant s}$ annihilates $\mathrm{Hom}_{\mathsf{T}}^*(C, D)$. Noting that $R^{\geqslant s}$ is an ideal of R, as $R^i = 0$ for $i < 0$, one gets the inclusion below:

$$\mathrm{supp}_R \mathrm{Hom}_{\mathsf{T}}^*(D, C) = \mathrm{supp}_R \mathrm{Hom}_{\mathsf{T}}^*(C, D) \subseteq \mathcal{V}(R^{\geqslant s}) = \mathcal{V}(R^{\geqslant 1}) \,.$$

The equality on the left is by Theorem 4.24, while the one on the right holds because $R^{\geqslant 1}$ and $R^{\geqslant s}$ have the same radical. The desired vanishing is thus a consequence of the exercise from the previous paragraph.   □

In [12, Section 5] we present variations of Theorem 4.24 where, for example, D need not be compact. Using these one can give another proof of the classification of thick subcategories of compact objects, Theorem 4.23.

4.2.3   Tensor triangulated categories

Let T be a tensor triangulated category with a canonical R-action. We assume that T is stratified as a tensor triangulated category, so that for each $\mathfrak{p} \in \mathrm{supp}_R \mathsf{T}$, the subcategory $\Gamma_{\mathfrak{p}} \mathsf{T}$, which is tensor ideal, contains no non-zero tensor ideal localising subcategories. There is then a *tensor product theorem* for support:

Theorem 4.26. *For any objects X, Y in T there is an equality*

$$\mathrm{supp}_R(X \otimes Y) = \mathrm{supp}_R X \cap \mathrm{supp}_R Y \,.$$

Compare this identity with the one in Theorem 4.24. For compact objects, these results can be deduced from each other, using $\mathcal{H}om(C, D) \cong \mathcal{H}om(C, \mathbb{1}) \otimes D$. Thus, Theorem 4.24 may be viewed as an analogue of the tensor product theorem for categories that admit no tensor product.

Proof. For each $\mathfrak{p} \in \mathrm{Spec}\, R$ there are isomorphisms

$$\Gamma_{\mathfrak{p}}(X \otimes Y) \cong \Gamma_{\mathfrak{p}} X \otimes \Gamma_{\mathfrak{p}} Y \cong \Gamma_{\mathfrak{p}} X \otimes Y \,.$$

These follow from the fact that $\Gamma_{\mathfrak{p}}(-) \cong \Gamma_{\mathfrak{p}} \mathbb{1} \otimes (-)$; see Section 4.1.4. They yield an inclusion

$$\mathrm{supp}_R(X \otimes Y) \subseteq \mathrm{supp}_R X \cap \mathrm{supp}_R Y \,.$$

As to the reverse inclusion: When $\Gamma_{\mathfrak{p}} X \neq 0$, the stratification condition implies that $\Gamma_{\mathfrak{p}} \mathbb{1}$ is in $\mathrm{Loc}_{\mathsf{T}}^{\otimes}(\Gamma_{\mathfrak{p}} X)$, so that $\Gamma_{\mathfrak{p}} Y$ is in $\mathrm{Loc}^{\otimes}(\Gamma_{\mathfrak{p}} X \otimes Y)$. Thus if $\Gamma_{\mathfrak{p}} Y \neq 0$ also holds, then $\Gamma_{\mathfrak{p}}(X \otimes Y) \neq 0$. This completes the proof.   □

4.3 The Klein four group

In this lecture we make some explicit calculations. We show how to compute the local cohomology functors using homotopy colimits. Then we illustrate this method by looking at a specific example. We use the Klein four group because its group algebra is of tame representation type. Thus one has a complete classification of all finite-dimensional representations.

4.3.1 Homotopy colimits

Let T be an R-linear triangulated category, and let $X_1 \xrightarrow{f_1} X_2 \xrightarrow{f_2} X_3 \xrightarrow{f_3} \cdots$ be a sequence of morphisms in T. Its *homotopy colimit*, denoted $\operatorname{hocolim} X_n$, is defined by an exact triangle

$$\bigoplus_{n \geqslant 1} X_n \xrightarrow{\theta} \bigoplus_{n \geqslant 1} X_n \longrightarrow \operatorname{hocolim} X_n \longrightarrow$$

where θ is the map $(\mathrm{id} - f_n)$; see [18].

Now fix a homogeneous element $r \in R$ of degree d. For each X in T and each integer n set $X_n = \Sigma^{nd}X$ and consider the commuting diagram

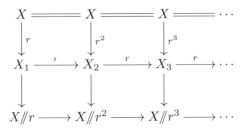

where each vertical sequence is given by the exact triangle defining $X/\!/r^n$, and the morphisms in the last row are the (non-canonical) ones induced by the commutativity of the upper squares. The gist of the next result is that the homotopy colimits of the horizontal sequences in the diagram compute $L_{\mathcal{V}(r)}X$ and $\Gamma_{\mathcal{V}(r)}X$. In [12, Proposition 2.9] we prove:

Proposition 4.27. *Let $r \in R$ be a homogeneous element of degree d. For each X in T the adjunction morphisms $X \to L_{\mathcal{V}(r)}X$ and $\Gamma_{\mathcal{V}(r)}X \to X$ induce isomorphisms*

$$\operatorname{hocolim} X_n \xrightarrow{\sim} L_{\mathcal{V}(r)}X \quad and \quad \operatorname{hocolim} \Sigma^{-1}(X/\!/r^n) \xrightarrow{\sim} \Gamma_{\mathcal{V}(r)}X. \qquad \square$$

From this, a standard argument based on an induction on the number of generators of $\mathfrak{a}$ yields the result below; see [12, Proposition 2.11].

Proposition 4.28. *For each ideal $\mathfrak{a}$ of $\operatorname{Spec} R$ and object $X \in \mathsf{T}$ one has*

$$\operatorname{Loc}_{\mathsf{T}}(\Gamma_{\mathcal{V}(\mathfrak{a})}X) = \operatorname{Loc}_{\mathsf{T}}(X/\!/\mathfrak{a}). \qquad \square$$

Note that the left-hand side depends only on the radical of $\mathfrak{a}$; cf. Lemma 3.11.

4.3.2   Representations of the Klein four group

We describe the representation theory of the Klein four group, and illustrate the abstract notions of the seminar using this example.

Let $G = \langle g_1, g_2 \rangle \cong \mathbb{Z}/2 \times \mathbb{Z}/2$ and let k be an algebraically closed field of characteristic 2. Let kG be the group algebra of G over k, and let $x_1 = g_1 - 1$, $x_2 = g_2 - 1$ as elements of kG. Then $x_1^2 = x_2^2 = 0$, and we have

$$kG = k[x_1, x_2]/(x_1^2, x_2^2).$$

We describe kG-modules by diagrams in which the vertices represent basis elements as a k-vector space, and an edge

indicates that $x_i a = b$. If there is no edge labelled x_i in the downwards direction from a vertex, then x_i sends the corresponding basis vector to zero. For example, the group algebra kG has the following diagram:

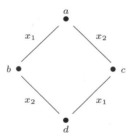

As a vector space, $kG = ka \oplus kb \oplus kc \oplus kd$. We have $\mathrm{rad}^2\, kG = \mathrm{soc}\, kG = kd$, $\mathrm{rad}\, kG = \mathrm{soc}^2\, kG = kb \oplus kc \oplus kd$.

Here are the diagrams for the syzygies of the trivial module:

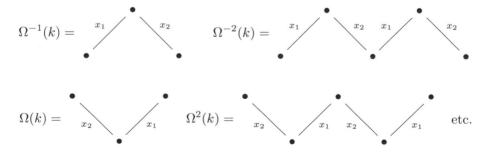

For each non-negative integer we have

$$\mathrm{Ext}^n_{kG}(k, k) \cong \underline{\mathrm{Hom}}_{kG}(k, \Omega^{-n}(k))$$

and so $\dim_k \mathrm{Ext}^n_{kG}(k, k) = n + 1$. In fact, the full cohomology algebra is

$$H^*(G, k) = \mathrm{Ext}^*_{kG}(k, k) = k[\zeta_1, \zeta_2]$$

with $\deg(\zeta_1) = \deg(\zeta_2) = 1$; see also Proposition 1.36.

Pick a non-zero element $r = r_1\zeta_1 + r_2\zeta_2$ in $H^1(G, k)$ (with $r_i \in k$). Then each power r^j gives us an injective map from k to $\Omega^{-j}(k)$ whose cokernel we denote by L_{r^j}. Thus we get a commutative diagram with exact columns

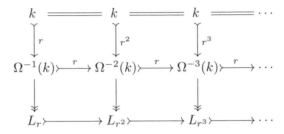

Thus, in $\mathsf{StMod}\, kG$ the module L_{r^j} is the Koszul object $k /\!\!/ r^j$, for each $j \geq 1$. In the case where $r = \zeta_1$, for example the diagrams are as follows:

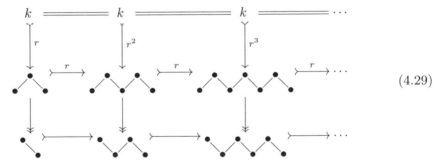

(4.29)

In this diagram, the k in the top row injects as the leftmost vertex in the modules in the middle row. The case of a general element r can be reduced to this one by applying automorphisms of the group algebra.

Remark 4.30. We can now describe the classification of the finite-dimensional indecomposable kG-modules; for details see [4, §4.3]. They come in three types:

(1) The group algebra kG itself.

(2) For each $n \in \mathbb{Z}$, the module $\Omega^n(k)$.

(3) For each one-dimensional subspace of $H^1(G, k)$, choose an element r; then for each integer $n \geq 1$ there is a module L_{r^n}. Thus we get a family of modules indexed by $\mathbb{P}^1(k) \times \mathbb{Z}_{\geq 1}$.

The modules in the third family for a particular choice of r, namely those appearing in the bottom row of diagram (4.29), form a tube in the Auslander–Reiten quiver of kG. The maps going up the tube are the maps in the bottom row of the diagram.

For larger elementary abelian groups the representation type is wild, and we cannot write down such a classification. Even for the Klein four group, the infinite-dimensional modules cannot be classified.

Next we compute the local cohomology functor $\Gamma_{\mathfrak{p}}\colon \mathsf{StMod}\,kG \to \mathsf{StMod}\,kG$ for each homogeneous prime ideal $\mathfrak{p}$ of $H^*(G,k)$. Observe that $\Gamma_{\mathfrak{m}} = 0$ for the unique maximal ideal $\mathfrak{m} = H^+(G,k)$. Now the homogeneous non-maximal prime ideals of $H^*(G,k)$ are the zero ideal $\mathfrak{n} = (0)$, and the principal ideals (r), one for each one-dimensional subspace of $H^1(G,k)$. Observe that

$$\Gamma_{(r)} = \Gamma_{\mathcal{V}(r)} \quad \text{and} \quad \Gamma_{\mathfrak{n}} = L_{\mathcal{Z}(\mathfrak{n})},$$

since each ideal (r) is maximal among all non-maximal ideals, and since $\mathfrak{n}$ is the unique minimal ideal; see Exercise 8 for Thursday.

The finite-dimensional indecomposables in

$$\Gamma_{\mathcal{V}(r)}\,\mathsf{StMod}(kG) \subseteq \mathsf{StMod}(kG)$$

are precisely the modules L_{r^n}, $n \geq 1$. For $r = \zeta_1$, the module $L_{\mathcal{V}(r)}(k)$ is the colimit of the modules in the middle row of diagram (4.29), while the module $\Gamma_{\mathcal{V}(r)}(k)$ is the colimit of the modules in the bottom row of the diagram. We can draw diagrams of these infinite-dimensional modules as follows:

$$L_{\mathcal{V}(r)}(k) = \ \nearrow\!\!\diagdown\!\!\nearrow\!\!\diagdown\!\!\nearrow\!\!\diagdown\!\!\nearrow \cdots$$

$$\Gamma_{\mathcal{V}(r)}(k) = \ \diagdown\!\!\nearrow\!\!\diagdown\!\!\nearrow\!\!\diagdown\!\!\nearrow\!\!\diagdown \cdots$$

The colimit of the vertical exact sequences in diagram (4.29) is the exact sequence

$$0 \to k \to L_{\mathcal{V}(r)}(k) \to \Gamma_{\mathcal{V}(r)}(k) \to 0$$

given by the inclusion of the left-hand vertex in $L_{\mathcal{V}(r)}(k)$.

The remaining prime to deal with is $\mathfrak{n}$, the zero ideal. The module $L_{\mathcal{Z}(\mathfrak{n})}(k)$ can be described as follows. Let $k(t)$ be the field of rational functions in one variable, regarded as an infinite-dimensional vector space over k. Then $L_{\mathcal{Z}(\mathfrak{n})}(k)$ is the module whose underlying vector space is a direct sum of two copies of $k(t)$, with G-action given by

$$g_1 \mapsto \begin{pmatrix} I & 0 \\ I & I \end{pmatrix}, \qquad g_2 \mapsto \begin{pmatrix} I & 0 \\ t.I & I \end{pmatrix}.$$

where I is the identity endomorphism of $k(t)$ and $t.I$ is the endomorphism of $k(t)$ given by multiplication by t. There is a map from k to $L_{\mathcal{Z}(\mathfrak{n})}(k)$ sending the identity to the vector $\begin{pmatrix} 0 \\ 1 \end{pmatrix}$, and the cokernel is $\Gamma_{\mathcal{Z}(\mathfrak{n})}(k)$. This gives the exact sequence

$$0 \to k \to L_{\mathcal{Z}(\mathfrak{n})}(k) \to \Gamma_{\mathcal{Z}(\mathfrak{n})}(k) \to 0.$$

Note that the modules $\Gamma_{\mathcal{V}}(k)$ are all periodic of period 1 in this example, so that a short exact sequence

$$0 \to k \to L_{\mathcal{V}}(k) \to \Gamma_{\mathcal{V}}(k) \to 0$$

gives rise to a triangle

$$\Gamma_{\mathcal{V}}(k) \to k \to L_{\mathcal{V}}(k) \to$$

in the stable module category, which is the localisation triangle for $\mathcal{V}$.

We can now draw a diagrammatic representation of the set of thick subcategories of $\mathsf{D}^b(\mathrm{mod}(kG))$.

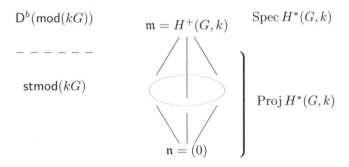

In this diagram, the dotted circle in the middle represents the set of closed points of the projective line over k, indexing the choices of (r). The part below the top vertex represents $\mathsf{stmod}(kG)$ while the whole diagram represents $\mathsf{D}^b(\mathrm{mod}(kG))$.

4.4 Exercises

(1) Let T be a compactly generated triangulated category. Given any class C of compact objects, prove that there exists a localisation functor $L \colon \mathsf{T} \to \mathsf{T}$ such that $\mathrm{Ker}\, L = \mathrm{Loc}(\mathsf{C})$.

Hint: Use Brown representability to show that the inclusion $\mathrm{Loc}(\mathsf{C}) \to \mathsf{T}$ admits a right adjoint.

(2) Let $A = \left[\begin{smallmatrix} k & k \\ 0 & k \end{smallmatrix}\right]$ be the algebra of 2×2 upper triangular matrices over a field k and let T denote the derived category of all A-modules. Up to isomorphism, there are precisely two indecomposable projective A-modules:

$$P_1 = \begin{bmatrix} k & 0 \\ 0 & 0 \end{bmatrix} \quad \text{and} \quad P_2 = \begin{bmatrix} 0 & k \\ 0 & k \end{bmatrix}$$

satisfying $\mathrm{Hom}_A(P_1, P_2) \neq 0$ and $\mathrm{Hom}_A(P_2, P_1) = 0$. For $i = 1, 2$ let L_i denote the localisation functor such that the L_i-acyclic objects form the smallest localising subcategory containing P_i, viewed as a complex concentrated in degree zero. Show that $L_1 L_2 \neq L_2 L_1$.

(3) Let T be a an R-linear triangulated category. Each specialisation closed subset of $\operatorname{Spec} R$ gives rise to a recollement of T in the sense of Section 3.3. Describe all 6 functors explicitly, and compare it with the prototype of a recollement given in [3, §1.4].

(4) Let k be a field, $A = k[x]/x(x-1)$, and T its derived category $\mathsf{D}(A)$. Then T is A-linear, hence also k-linear, via restriction along the homomorphism $k \to A$. Prove that T has four localising subcategories. Hence T cannot be stratified by the k-action. It is however stratified by the A-action; this is a special case of Neeman's theorem, but can be verified directly.

(5) Prove that $\operatorname{Ext}_{\mathbb{Z}}^*(\mathbb{Q}, \mathbb{Z}) = 0$. Thus the 'orthogonality' relation:

$$\operatorname{Supp}_{\mathbb{Z}} \operatorname{Hom}^*_{\mathsf{D}(\mathbb{Z})}(C, Y) = \operatorname{Supp}_{\mathbb{Z}} C \cap \operatorname{Supp}_{\mathbb{Z}} Y$$

may fail when C is not compact; i.e., when $H^*(C)$ is not finitely generated.

Another example to bear in mind is a complete local ring A, with maximal ideal $\mathfrak{m}$. With E denoting the injective hull of $A/\mathfrak{m}$ one has

$$\operatorname{Supp}_A \operatorname{Hom}^*_{\mathsf{D}(A)}(E, E) = \operatorname{Supp}_A \operatorname{Hom}_A(E, E) = \operatorname{Supp}_A A = \operatorname{Spec} A.$$

On the other hand, $\operatorname{Supp}_A E = \{\mathfrak{m}\}$.

(6) Prove that $\operatorname{supp}_R \mathsf{T} = \operatorname{supp}_R \mathsf{T}^{\mathsf{c}}$.

(7) Let T be an R-linear triangulated category. Prove that there are equalities

$$\{\mathfrak{p} \in \operatorname{Spec} R \mid X_{\mathfrak{p}} \neq 0\} = \bigcup_{C \in \mathsf{T}^{\mathsf{c}}} \operatorname{Supp}_R \operatorname{Hom}^*_{\mathsf{T}}(C, X) = \operatorname{cl}(\operatorname{supp}_R X)$$

for each X in T. The set on the left is what we denoted $\operatorname{Supp}_R X$.

(8) Let $\mathfrak{p}$ be a homogeneous prime ideal in R. Prove that $\Gamma_{\mathfrak{p}} \cong \Gamma_{\mathcal{V}(\mathfrak{p})}$ when $\mathfrak{p}$ is maximal, with respect to inclusion, in $\operatorname{supp}_R \mathsf{T}$, and that $\Gamma_{\mathfrak{p}} \cong L_{\mathcal{Z}(\mathfrak{p})}$ when $\mathfrak{p}$ is minimal in $\operatorname{supp}_R \mathsf{T}$.

(9) For any object X in T and homogeneous ideal $\mathfrak{a}$ in R, prove that

$$\operatorname{supp}_R(X /\!\!/ \mathfrak{a}) = \mathcal{V}(\mathfrak{a}) \cap \operatorname{supp}_R X \,;$$

cf. Proposition 2.5. Using this, prove that any closed subset of $\operatorname{supp}_R \mathsf{T}$ is realizable as the support of a compact object in T.

(10) Let M and N be kG-modules. Recall (or prove) that there is an isomorphism

$$\operatorname{Ext}^*_{kG}(M, N) \cong H^*(G, \operatorname{Hom}_k(M, N)),$$

compatible with the $H^*(G, k)$-actions. Assume now that M and N are finitely generated and prove that there is an isomorphism of kG-modules:

$$\operatorname{Hom}_k(\operatorname{Hom}_k(M, N), k) \cong \operatorname{Hom}_k(N, M)$$

Conclude that the support of $\operatorname{Ext}^*_{kG}(M, N)$ and $\operatorname{Ext}^*_{kG}(N, M)$, as modules over $H^*(G, k)$, coincide. Compare with Theorem 4.24.

5 Friday

In this last chapter we put together the various ideas we have been developing in the week's lectures. The goal, as has been stated often enough, is a classification of the localising subcategories of the stable module category of a finite group, over a field of characteristic p. The strategy of the proof was described in Section 4.1.6, and we begin this chapter at the last step, which is also where the whole story begins, namely, Neeman's classification of the localising subcategories of the derived category of a commutative noetherian ring. Using a (version of) this result, and a variation of the Bernstein-Gelfand-Gelfand correspondence, we explain how to tackle the case of the homotopy category of complexes of injective modules over an elementary abelian two group. This is the content of Section 5.2. Finally, in the last section, we use Quillen's results to describe how to pass from arbitrary groups to elementary abelian ones. If the dust settles down, the reader should be able to see a fairly complete proof of our main results for the case $p = 2$.

5.1   Localising subcategories of $\mathsf{D}(A)$

In this lecture A is a commutative noetherian ring and D the derived category of the category of A-modules. Recall that D is a triangulated category admitting set-indexed coproducts. Its compact objects are:

$$\mathsf{D}^{\mathrm{c}} = \mathrm{Thick}(A)\,,$$

the set of perfect complexes of A-modules; see Theorem 2.2. Moreover, one has

$$\mathrm{Loc}_A(A) = \mathsf{D}\,,$$

so D is a compactly generated triangulated category, with A a compact generator. The category D is also A-linear, with structure homomorphisms

$$A \to \mathrm{Hom}^*_{\mathsf{D}}(M, M)$$

given by left multiplication. As A is a compact generator and $\mathrm{Hom}^*_{\mathsf{D}}(A, A) \cong A$, it follows that the A-linear category D is noetherian. Note that $\mathrm{supp}_A \mathsf{D} = \mathrm{Spec}\, A$.

This lecture is devoted to proving the following result:

Theorem 5.1. *The triangulated category* D *is stratified by* A.

A proof is given later in this lecture. For now, we record some direct corollaries. The one below is by Theorem 4.19, and is one of the main results in [44].

Corollary 5.2. *There is bijection between localising subcategories of* D *and subsets of* $\operatorname{Spec} A$. $\qquad\square$

The next result is by Theorem 4.23, since D is a noetherian category.

Corollary 5.3. *There is bijection between the thick subcategories of* $\operatorname{Thick}(A)$ *and specialisation closed subsets of* $\operatorname{Spec} A$. $\qquad\square$

5.1.1 Local cohomology and support

Next we relate the functors $\Gamma_{\mathcal{V}}$ and $L_{\mathcal{V}}$ to familiar functors from commutative algebra. What follows is a summary of results from [10, Section 9].

To begin with, note that for each M in D there is an isomorphism of A-modules

$$\operatorname{Hom}^*_{\mathsf{D}}(A, M) \cong H^*(M)$$

For any $\mathfrak{p} \in \operatorname{Spec} A$, there are natural isomorphisms

$$L_{\mathcal{Z}(\mathfrak{p})}(M) \cong M_{\mathfrak{p}},$$

where $M_{\mathfrak{p}}$ denotes the usual localisation of the complex M at $\mathfrak{p}$. In other words, the functor $M \mapsto M_{\mathfrak{p}}$ is the localisation with respect to $\mathcal{Z}(\mathfrak{p})$.

For each ideal $\mathfrak{a}$ in A, the functor $\Gamma_{\mathcal{V}(\mathfrak{a})}$ is the derived functor of the $\mathfrak{a}$-*torsion functor*, that associates to each A-module M the submodule

$$\{m \in M \mid \mathfrak{a}^n m = 0 \text{ for some } n \geq 0\}.$$

It follows that, for any M in D, the cohomology of $\Gamma_{\mathcal{V}(\mathfrak{a})}(M)$ is the *local cohomology* of M with support in $\mathfrak{a}$, in the sense of Grothendieck:

$$H^*(\Gamma_{\mathcal{V}(\mathfrak{a})}M) \cong H^*_{\mathfrak{a}}(M) \quad \text{for each } M \in \mathsf{D}.$$

In particular, for each $\mathfrak{p} \in \operatorname{Spec} A$ there is an isomorphism $H^*(\Gamma_{\mathfrak{p}}M) \cong H^*_{\mathfrak{p}A_{\mathfrak{p}}}(M_{\mathfrak{p}})$, so that the support of M is computed as:

$$\operatorname{supp}_A M = \{\mathfrak{p} \in \operatorname{Spec} A \mid H^*_{\mathfrak{p}A_{\mathfrak{p}}}(M_{\mathfrak{p}}) \neq 0\}.$$

Compare this with other methods for computing support given in Lemma A.14.

5.1.2   Proof of Theorem 5.1

The derived tensor product $- \otimes_A^{\mathbf{L}} -$ on D endows it with a structure of a tensor triangulated category, with unit A. Since A is a compact generator for D, it follows from Theorem 4.9 that the local global principle holds for D. It thus remains to verify stratification condition (S2) from Section 4.2.

Fix $\mathfrak{p} \in \operatorname{Spec} A$ and set $k(\mathfrak{p}) = A_\mathfrak{p}/\mathfrak{p}A_\mathfrak{p}$, the residue field of A at $\mathfrak{p}$. Note that $k(\mathfrak{p})$ is $\mathfrak{p}$-local and $\mathfrak{p}$-torsion and hence is in $\Gamma_\mathfrak{p}$D, by Corollary 3.9. In particular, the latter category is non-empty, and to prove that $\Gamma_\mathfrak{p}$D is minimal, it thus suffices to prove:

$$\operatorname{Loc}_{\mathsf{D}}(M) = \operatorname{Loc}_{\mathsf{D}}(k(\mathfrak{p})) \quad \text{for any } M \in \Gamma_\mathfrak{p}\mathsf{D} \text{ with } H^*(M) \neq 0.$$

We reduce to the case where A is a local ring, with maximal ideal $\mathfrak{p}$, as follows: The homomorphism of rings $A \to A_\mathfrak{p}$ gives a restriction functor $\iota \colon \mathsf{D}(A_\mathfrak{p}) \to \mathsf{D}(A)$. Note that there is an isomorphism $M \cong \iota(M_\mathfrak{p})$ in D, because M is $\mathfrak{p}$-local. Since ι is compatible with coproducts, and maps $k(\mathfrak{p})$ (viewed as an $A_\mathfrak{p}$-module) to $k(\mathfrak{p})$, it suffices to prove

$$\operatorname{Loc}_{\mathsf{D}(A_\mathfrak{p})}(M_\mathfrak{p}) = \operatorname{Loc}_{\mathsf{D}(A_\mathfrak{p})}(k(\mathfrak{p})).$$

Observe that $M_\mathfrak{p}$ and $k(\mathfrak{p})$ are local and torsion with respect to the ideal $\mathfrak{p}A_\mathfrak{p} \subset A_\mathfrak{p}$. Hence replacing A and M by $A_\mathfrak{p}$ and $M_\mathfrak{p}$, we may assume A is local, with maximal ideal $\mathfrak{p}$, and set $k = A/\mathfrak{p}$. Since $\Gamma_\mathfrak{p} = \Gamma_{\mathcal{V}(\mathfrak{p})}$, by Exercise 8 from Thursday's lecture, we then need to prove that

$$\operatorname{Loc}_{\mathsf{D}}(M) = \operatorname{Loc}_{\mathsf{D}}(k) \quad \text{for each } M \in \Gamma_{\mathcal{V}(\mathfrak{p})}\mathsf{D}.$$

First we tackle the case $M = \Gamma_{\mathcal{V}(\mathfrak{p})}A$. We claim k is in $\operatorname{Loc}_{\mathsf{D}}(\Gamma_{\mathcal{V}(\mathfrak{p})}A)$. Indeed, since localising subcategories of D are tensor closed, one has:

$$(\Gamma_{\mathcal{V}(\mathfrak{p})}A \otimes_A^{\mathbf{L}} k) \in \operatorname{Loc}_{\mathsf{D}}(\Gamma_{\mathcal{V}(\mathfrak{p})}A).$$

Now note that there are isomorphisms $k \cong \Gamma_{\mathcal{V}(\mathfrak{p})}k \cong (\Gamma_{\mathcal{V}(\mathfrak{p})}A \otimes_A^{\mathbf{L}} k)$. Therefore

$$\operatorname{Loc}_{\mathsf{D}}(k) \subseteq \operatorname{Loc}_{\mathsf{D}}(\Gamma_{\mathcal{V}(\mathfrak{p})}A).$$

We verify the reverse inclusion by verifying the follows claims:

$$\Gamma_{\mathcal{V}(\mathfrak{p})}A \in \operatorname{Loc}_{\mathsf{D}}(A/\!\!/\mathfrak{p}) \quad \text{and} \quad A/\!\!/\mathfrak{p} \in \operatorname{Thick}_{\mathsf{D}}(k).$$

The first inclusion is from Proposition 4.28. As to the second: Since A is finitely generated as an A-module and $A/\!\!/\mathfrak{p}$ is in the thick subcategory generated by A, one obtains that the A-module $H^*(A/\!\!/\mathfrak{p})$ is finitely generated. It is also $\mathfrak{p}$-torsion, by Proposition 3.12, and hence of finite length. Since A is local, with residue field k, it follows that $A/\!\!/\mathfrak{p}$ is in the thick subcategory generated by k; see Exercise 6.

To sum up: $\mathrm{Loc}_\mathsf{D}(\Gamma_{\mathcal{V}(\mathfrak{p})}A) = \mathrm{Loc}_\mathsf{D}(k)$. Since localising subcategories of D are tensor closed, and the exact functor $-\otimes_A^\mathbf{L} M$ preserves coproducts, the preceding equality yields the second equality below:

$$\mathrm{Loc}_\mathsf{D}(M) = \mathrm{Loc}_\mathsf{D}(\Gamma_{\mathcal{V}(\mathfrak{p})}A \otimes_A^\mathbf{L} M) = \mathrm{Loc}_\mathsf{D}(k \otimes_A^\mathbf{L} M)\,.$$

The first equality holds because M is in $\Gamma_{\mathcal{V}(\mathfrak{p})}\mathsf{D}$; see Section 4.1.4. Since $H^*(M)$ is non-zero, one gets in particular that $H^*(k \otimes_A^\mathbf{L} M)$ is non-zero as well. Since k is a field, in D there is an isomorphism

$$k \otimes_A^\mathbf{L} M \cong H^*(k \otimes_A^\mathbf{L} M)\,.$$

Since $H^*(k \otimes_A^\mathbf{L} M)$ is a non-zero k-vector space, it can build k and vice-versa, in $\mathsf{D}(k)$ and hence also in D.

This completes the proof of the theorem.   □

5.2   Elementary abelian 2-groups

Fix a prime number p, and let $E = \langle g_1, \ldots, g_r \rangle$ be an elementary abelian p-group of rank r. We consider $\mathsf{K}(\mathsf{Inj}\,kE)$, the homotopy category of complexes of injective (which in this case is the same as free) kE-modules. Recall that the tensor product over k, with diagonal kE-action, induces on $\mathsf{K}(\mathsf{Inj}\,kE)$ a structure of a tensor triangulated category, and that there is a canonical action of $H^*(E, k)$ on it.

As explained in Section 4.1.6, one of the main steps in [10] is a proof of the statement that $\mathsf{K}(\mathsf{Inj}\,kE)$ is stratified by $H^*(E, k)$. In this lecture, we prove this result in the special case where $p = 2$, namely:

Theorem 5.4. *Let k be a field of characteristic 2 and E an elementary abelian 2-group. Then $\mathsf{K}(\mathsf{Inj}\,kE)$ is stratified by the canonical action of $H^*(E, k)$.*

The proof of this theorem is given in Section 5.2.3. It uses a variation, and enhancement, of the classical Bernstein–Gelfand–Gelfand correspondence [17]; see Remark 5.6 for a detailed comparison. The statement, and the proof, of this latter result requires a foray into some homological algebra of differential graded modules over differential graded algebras.

5.2.1   Differential graded algebras

Let A be differential graded algebra. This means that A is a graded algebra with a differential satisfying the Leibniz rule:

$$d(ab) = d(a)\,b + (-1)^{|a|}a\,d(b)$$

for all homogeneous elements a, b in A. In the same vein, a differential graded module M over such a differential graded algebra A is a graded A-module with a differential such that the multiplication satisfies the appropriate Leibniz rule.

Any graded algebra A can be viewed as a differential graded algebra with zero differential; most of the differential graded algebras we encounter in the discussion below will be of this nature. In this case, any graded A-module is a differential graded A-module with zero differential. However, the category of differential graded A-modules is typically much larger, as should be clear soon. A ring is a differential graded algebra concentrated in degree 0, and then a differential graded module is nothing more, and nothing less, than a complex of modules.

Let A be a differential graded algebra. A morphism $f \colon M \to N$ of differential graded A-modules is a homomorphism of graded A-modules that is at the same time a morphism of complexes. Differential graded A-modules and their morphisms form an abelian category. The usual notion of homotopy (for complexes of modules over a ring) carries over to differential graded modules, and one can form the associated homotopy category. Inverting the quasi-isomorphisms there gives the *derived category* $\mathsf{D}(A)$ of differential graded A-modules. It is triangulated and the differential graded module A is a compact object and a generator; see [40].

5.2.2   A BGG correspondence

Let k be a field of characteristic 2. Let E be an elementary abelian 2-group and kE its group algebra. Thus $E \cong (\mathbb{Z}/2)^r$ and, setting $z_i = g_i - 1$, there is an isomorphism of k-algebras

$$kE \cong k[z_1, \ldots, z_r]/(z_1^2, \ldots, z_r^2) \,.$$

Let $S = k[x_1, \ldots, x_r]$ be a graded polynomial algebra over k where $|x_i| = 1$ for each i. We view it as a differential graded algebra with zero differential. Our goal is to construct an explicit equivalence between the triangulated categories $\mathsf{K}(\mathsf{Inj}\, kE)$ and $\mathsf{D}(S)$, the derived category of differential graded modules over S. To this end, we mimic the approach in [2, Section 7].

The graded k-algebra $kE \otimes_k S$, with component in degree i being $kE \otimes_k S^i$ and product defined by

$$(a \otimes s) \cdot (b \otimes t) = ab \otimes st \,,$$

is commutative. Again we view it as a differential graded algebra with zero differential, and consider in it the element of degree 1:

$$\delta = \sum_{i=1}^{r} z_i \otimes_k x_i \,.$$

It is easy to verify that $\delta^2 = 0$; keep in mind that the characteristic of k is 2. In what follows J denotes the differential graded module over $kE \otimes_k S$ with underlying graded module and differential given by

$$J = kE \otimes_k S \quad \text{and} \quad d(e) = \delta e \,.$$

Observe that since J is a differential graded module over $kE \otimes_k S$, for each differential graded module M over kE there is an induced structure of a differential graded S-module on $\mathrm{Hom}_{kE}(J, M)$. Moreover, the map

$$\zeta \colon S \to \mathrm{Hom}_{kE}(J, J)$$

induced by right multiplication is a morphism of differential graded algebras.

Theorem 5.5. *The map ζ is a quasi-isomorphism and the functor*

$$\mathrm{Hom}_{kE}(J, -) \colon \mathsf{K}(\mathsf{Inj}\, kE) \to \mathsf{D}(S)$$

is an equivalence of triangulated categories.

Proof. The main ingredients in the proof are summarised in the following claims:

Claim: As a complex of kE-modules, J consists of injectives and $J^i = 0$ for $i < 0$.

Indeed, the degree i component of J is $kE \otimes_k S^i$, which is isomorphic to a direct sum of $\binom{r+i-1}{i}$ copies of kE. The claim follows, since kE is self-injective.

Set $w = z_1 \cdots z_r$; this is a non-zero element in the socle of kE, and hence generates it. It is easy to verify that the map of graded k-vector spaces

$$\eta \colon k \to J \quad \text{defined by } \eta(1) = w \otimes 1$$

is a morphism of differential graded kE-modules.

Claim: $\eta \colon k \to J$ is an injective resolution of k, as a kE-module.

Given the previous claim, and that η is kE-linear, it remains to prove that η is a quasi-isomorphism. We have to prove that $H^*(\eta)$ is an isomorphism, and in verifying this we can, and will, ignore the kE-module structures on k and J.

When $r = 1$, complex J can be identified with the complex

$$0 \to k[z]/(z^2) \xrightarrow{z} k[z]/(z^2) \xrightarrow{z} \cdots$$

of $k[z]/(z^2)$-modules. It is easy to verify that the only non-zero cohomology in this complex is in degree 0, where it is kw. Thus, η is a quasi-isomorphism.

For general r, writing $J(i)$ for the complex $k[z_i]/(z_i^2) \otimes_k k[x_i]$ one has a natural isomorphism

$$J \cong J(1) \otimes_k \cdots \otimes_k J(r)$$

of complex of k-vector spaces. With this, and setting $\eta(i)$ to be the map $k \to J(i)$ defined by $1 \mapsto z_i \otimes 1$, one gets an identification

$$\eta = \eta(1) \otimes_k \cdots \otimes_k \eta(r) \colon k \to J.$$

Since each $\eta(i)$ is a quasi-isomorphism, it follows from the Künneth formula that $H^*(\eta)$ is an isomorphism as well.

This completes the proof of the claim.

Now we are ready to prove that the map $\zeta\colon S \to \operatorname{Hom}_{kE}(J, J)$ induced by right multiplication is a quasi-isomorphism. In verifying that ζ is a quasi-isomorphism we can now ignore the S-module structures. The composite morphism of k-vector spaces

$$S \xrightarrow{\ \zeta\ } \operatorname{Hom}_{kE}(J, J) \xrightarrow{\ \operatorname{Hom}_{kE}(\eta, J)\ } \operatorname{Hom}_{kE}(k, J)$$

sends an element $s \in S$ to the map $1 \mapsto w \otimes s$, where $w = z_1 \ldots z_r$. It is easy to see that the composite is an isomorphism of complexes. It remains to note that $\operatorname{Hom}_{kE}(\eta, J)$ is a quasi-isomorphism: η is a quasi-isomorphism of differential graded modules over kE and $\operatorname{Hom}_{kE}(-, J)$ preserves quasi-isomorphisms, because J is a complex of injective kE-modules with $J^i = 0$ for $i < 0$.

Armed with these facts about J, the proof can be wrapped up as follows: It is easy to verify that the functor $\operatorname{Hom}_{kE}(J, -)$ from $\mathsf{K}(\operatorname{Inj} kE)$ to $\mathsf{D}(S)$ is exact. Since J is an injective resolution of k, it is compact and generates the triangulated category $\mathsf{K}(\operatorname{Inj} kE)$; see Theorem 3.17. It follows from this that the functor $\operatorname{Hom}_{kE}(J, -)$ preserves coproducts, since a quasi-isomorphism between differential graded S-modules is an isomorphism in $\mathsf{D}(S)$. As S is quasi-isomorphic to $\operatorname{Hom}_{kE}(J, J)$ and it is a compact generator for $\mathsf{D}(S)$, it remains to apply Exercise 23 in Chapter 1 to conclude that $\operatorname{Hom}_{kE}(J, -)$ is an equivalence. □

Remark 5.6. Let k be a field, Λ an exterior algebra on generators of degree 1, and S a polynomial algebra on generators of degree 1. Bernstein, Gelfand, and Gelfand [17] proved that there is an equivalence of triangulated categories

$$\mathsf{D}^b(\operatorname{grmod}\Lambda) \simeq \mathsf{D}^b(\operatorname{grmod} S).$$

The functor inducing the equivalence is similar to the one in Theorem 5.5.

Over a field of characteristic 2, the group algebra of an elementary abelian 2 group is an exterior algebra. However, the result above cannot be applied directly to our context. One point being that it deals with *graded modules* over an exterior algebra of degree 1, but this is not crucial. The main issue is that it deals with *bounded* derived categories, while for our purposes we need a statement at the level of the full homotopy category of complexes of injective modules; see Section 3.3.5 for the discrepancy between the two.

5.2.3   Proof of Theorem 5.4

We retain the notation from Section 5.2.2.

The basic idea of the proof is clear enough: $\mathsf{K}(\operatorname{Inj} kE)$ is equivalent to $\mathsf{D}(S)$, by Theorem 5.5, so it suffices to prove that the latter is stratified, and for that one invokes Theorem 5.1. Two issues need clarification.

One point is that Theorem 5.1 concerns commutative rings and not differential graded algebras. However, the argument given for *op. cit.* carries over with minor modifications to $\mathsf{D}(S)$; see [11, Theorem 5.2], and also [12, Theorem 8.1], for a statement that applies to any formal differential graded algebra whose homology is commutative and noetherian.

The second point is this: By the discussion above, $\mathsf{K}(\mathsf{Inj}\, kE)$ is stratified by the S-action induced by the equivalence of categories in Theorem 5.5. The latter result also gives that $H^*(E,k) \cong S$, as algebras; one then needs to check that the induced action of $H^*(E,k)$ on $\mathsf{K}(\mathsf{Inj}\, kE)$ is the diagonal action. In fact, much less is required, namely, that the local cohomology functors, $\Gamma_\mathfrak{p}$, defined by the two actions coincide, and this is easier to verify; see the proof of [11, Theorem 6.4].

For a systematic treatment of such "change-of-categories" results, which may serve to clarify the issue, see [13, Section 7].

5.3   Stratification for arbitrary finite groups

The goal for this lecture is as follows. Let G be a finite group and k a field of characteristic p. We shall assume that $\mathsf{K}(\mathsf{Inj}\, kE)$ is stratified by $H^*(E,k)$ for all elementary abelian subgroups E of G (for $p = 2$ this was proved in the previous lecture) and we shall prove that $\mathsf{K}(\mathsf{Inj}\, kG)$ is stratified by $H^*(G,k)$.

Our main tool is Quillen's Stratification Theorem; so we begin by examining Quillen's theorem in some detail.

5.3.1   Quillen Stratification

Quillen [48, 49] (1971) described the *maximal* ideal spectrum of $H^*(G,k)$ in terms of elementary abelian subgroups. We are really interested in *prime* ideals, but let us first describe Quillen's original theorem. Note that in a finitely generated graded commutative k-algebra such as $H^*(G,k)$ every prime ideal is the intersection of the maximal ideals containing it, by a version of Hilbert's Nullstellensatz.

Let G be a finite group, and k a field of characteristic p. If H is a subgroup of G, there is a *restriction map* $H^*(G,k) \to H^*(H,k)$ which is a ring homomorphism. Writing $V_G = \max H^*(G,k)$, this induces a map of varieties $V_H \to V_G$.

Theorem 5.7. *An element* $u \in H^*(G,k)$ *is nilpotent if and only if* $\mathrm{res}_{G,E}(u)$ *is nilpotent for every elementary abelian p-subgroup* $E \leq G$.   □

It thus makes sense to look at the product of the restriction maps

$$H^*(G,k) \to \prod_E H^*(E,k).$$

The theorem implies that the kernel of this map is nilpotent. What is the image?

If an element (u_E) is in the image, then

(1) for each conjugation $c_g \colon E' \to E$ in G, $c_g^*(u_E) = u_{E'}$ and

(2) for each inclusion $i \colon E' \to E$ in G, $i^*(u_E) = u_{E'}$.

Conversely, if (u_E) satisfies these conditions, Quillen showed that for some $t \geq 0$ the element $(u_E^{p^t})$ is in the image.

Definition 5.8. We define $\varprojlim_E H^*(E, k)$ to be the elements (u_E) of the direct product $\prod_E H^*(E, k)$ satisfying conditions (1) and (2) above.

Definition 5.9. A homomorphism $\phi\colon R \to S$ of graded commutative k-algebras is an *F-isomorphism* or *inseparable isogeny* if

(1) the kernel of ϕ consists of nilpotent elements, and

(2) for each $s \in S$ there exists $t \geq 0$ such that $s^{p^t} \in \mathrm{Im}\,\phi$.

We can now rephrase Quillen's theorem as follows:

Theorem 5.10. *The restriction maps induce an F-isomorphism*

$$H^*(G, k) \to \varprojlim_E H^*(E, k) \,. \qquad \square$$

Now if $\phi\colon R \to S$ is an F-isomorphism of finitely generated graded commutative k-algebras, then $\phi^*\colon \max S \to \max R$ is a *bijection*.

For the moment, let us assume that k is algebraically closed. Recalling that $V_G = \max H^*(G, k)$, this says that $\varinjlim_E V_E \to V_G$, as a map of varieties, is *bijective* (but not necessarily invertible!).

Here, $\varinjlim_E V_E$ is the quotient of the disjoint union of the V_E by the equivalence relation induced by the conjugations and inclusions.

5.3.2 A more concrete view

Let us continue for a while with the assumption that k is algebraically closed, a restriction which we shall later lift. Let us describe V_G more explicitly in terms of the elementary abelian p-subgroups. If E is an elementary abelian p-group of rank r, then $H^*(E, k)$ modulo nilpotents is a polynomial ring in r variables, and so

$$V_E = \max H^*(E, k) \cong \mathbb{A}^r(k),$$

affine space of dimension r.

If $E' \leq E$, then $\mathrm{res}^*_{E,E'}$ identifies $V_{E'}$ as a linear subspace of V_E determined by linear equations with coefficients in the ground field $\mathbb{F}_p$. Furthermore, each such linear subspace is the image of $\mathrm{res}^*_{E,E'}$ for a suitable $E' \leq E$.

Definition 5.11. We set $V_E^+ = V_E \setminus \bigcup_{E' < E} V_{E'}$, so that V_E^+ is obtained from V_E by removing all $(p^r - 1)/(p - 1)$ of the codimension 1 subspaces in V_E defined over $\mathbb{F}_p$. The set V_E^+ is a dense open subset of V_E.

We consider the following subvarieties of V_G:

$$V_{G,E} = \mathrm{res}^*_{G,E}(V_E) \qquad \text{and} \qquad V_{G,E}^+ = \mathrm{res}^*_{G,E}(V_E^+) \,.$$

Thus $V_{G,E}$ is a closed subvariety and $V_{G,E}^+$ is a locally closed subvariety.

Theorem 5.12. *The following statements hold*

(1) *V_G is the disjoint union of locally closed subsets $V_{G,E}^+$, one for each conjugacy class of elementary abelian subgroups E of G.*

(2) *$V_{G,E}^+$ is (F-isomorphic to) the quotient of the locally closed variety V_E^+ by the free action of $N_G(E)/C_G(E)$.*

(3) *The layers $V_{G,E}^+$ of V_G are glued together via the inclusions $E' < E$ in G.* □

5.3.3   Prime ideals

Now we discuss prime ideals in $H^*(G,k)$, and we drop the assumption that k is algebraically closed. The discussion below is based on [11, Section 9].

Write $\mathcal{V}_G$ for the (homogeneous) prime spectrum of $H^*(G,k)$. By Quillen's theorem, for each $\mathfrak{p} \in \mathcal{V}_G$ there exists an elementary abelian p-subgroup $E \leq G$ such that $\mathfrak{p}$ is in the image of $\mathrm{res}_{G,E}^*$. We say that $\mathfrak{p}$ *originates* in such an E if there does not exist a proper subgroup E' of E such that $\mathfrak{p}$ is in the image of $\mathrm{res}_{G,E'}^*$.

Theorem 5.13. *For each $\mathfrak{p} \in \mathcal{V}_G$, the pairs $(E,\mathfrak{q})$ where $\mathfrak{q} \in \mathcal{V}_E$, $\mathfrak{p} = \mathrm{res}_{G,E}^*(\mathfrak{q})$ and such that $\mathfrak{p}$ originates in E are all G-conjugate.* □

Warning 5.14. In contrast with part (2) of Theorem 5.12, if we fix $\mathfrak{p}$ and E such that $\mathfrak{p}$ originates in E, then $N_G(E)/C_G(E)$ acts transitively but not necessarily freely on the set of primes $\mathfrak{q} \in \mathcal{V}_E$ such that $\mathrm{res}_{G,E}^*(\mathfrak{q}) = \mathfrak{p}$.

Indeed, let E be the normal four group of the alternating group A_4, and let k be an algebraically closed field of characteristic 2. Let ω be a primitive cube root of unity in k. Write $H^*(E,k) = k[x,y]$ where x and y are defined over $\mathbb{F}_2$. Then the prime ideal $(x+\omega y)$ is invariant under the action of G but there is no inhomogeneous maximal ideal containing it that is fixed, because each point in the line is multiplied by ω or ω^2 when acted on by an element of A_4 not in E.

For any subgroup $H \leq G$, formal properties of restriction and induction prove the following:

Lemma 5.15. *Let $\mathfrak{p} \in \mathcal{V}_G$ and set $\mathcal{U} = (\mathrm{res}_{G,H}^*)^{-1}\{\mathfrak{p}\}$.*

(1) *For any X in $\mathsf{K}(\mathsf{Inj}\,kG)$, $(\Gamma_\mathfrak{p} X){\downarrow}_H \cong \bigoplus_{\mathfrak{q}\in\mathcal{U}} \Gamma_\mathfrak{q}(X{\downarrow}_H)$.*

(2) *For any Y in $\mathsf{K}(\mathsf{Inj}\,kH)$, $\Gamma_\mathfrak{p}(Y{\uparrow}^G) \cong \bigoplus_{\mathfrak{q}\in\mathcal{U}}(\Gamma_\mathfrak{q}Y){\uparrow}^G$.* □

Using this, we get:

Theorem 5.16. *If $X \in \mathsf{K}(\mathsf{Inj}\,kG)$ and $Y \in \mathsf{K}(\mathsf{Inj}\,kH)$, then*

(1) *$\mathcal{V}_G(X{\downarrow}_H{\uparrow}^G) \subseteq \mathcal{V}_G(X)$, and*

(2) *$\mathcal{V}_G(Y{\uparrow}^G) = \mathrm{res}_{G,H}^* \mathcal{V}_H(Y)$.* □

The following version of the subgroup theorem is for elementary abelian groups. The analogue for arbitrary finite groups also holds, and its proof makes use of this more limited version.

Theorem 5.17. *Let $E' \leq E$ be elementary abelian p-groups, $X \in \mathsf{K}(\mathsf{Inj}\, kE)$. Then*

$$\mathcal{V}_{E'}(X{\downarrow}_{E'}) = (\mathrm{res}^*_{E,E'})^{-1}\mathcal{V}_E(X).$$

Proof. Let $\mathfrak{q} \in \mathcal{V}_{E'}$, $\mathfrak{p} = \mathrm{res}^*_{E,E'}(\mathfrak{q})$. By the previous theorem we have

$$\mathcal{V}_E(\Gamma_{\mathfrak{q}}k{\uparrow}^E) = \{\mathfrak{p}\} = \mathcal{V}_E(\Gamma_{\mathfrak{p}}k),.$$

By the classification of localising subcategories for kE this implies that

$$\mathrm{Loc}(\Gamma_{\mathfrak{q}}k{\uparrow}^E) = \mathrm{Loc}(\Gamma_{\mathfrak{p}}k).$$

Thus

$$
\begin{aligned}
\Gamma_{\mathfrak{q}}(X{\downarrow}_{E'}) \neq 0 &\iff \Gamma_{\mathfrak{q}}k \otimes X{\downarrow}_{E'} \neq 0 \\
&\iff \Gamma_{\mathfrak{q}}k{\uparrow}^E \otimes X \neq 0 \\
&\iff \Gamma_{\mathfrak{p}}k \otimes X \neq 0 \\
&\iff \Gamma_{\mathfrak{p}}X \neq 0. \qquad\qquad \square
\end{aligned}
$$

5.3.4 Chouinard's Theorem for $\mathsf{K}(\mathsf{Inj}\, kG)$

The next ingredient in the proof of the classification theorem is a version of Chouinard's Theorem 1.7 for the category $\mathsf{K}(\mathsf{Inj}\, kG)$.

Theorem 5.18. *An object X in $\mathsf{K}(\mathsf{Inj}\, kG)$ is zero if and only if $X{\downarrow}_E$ is zero for every elementary abelian p-subgroup $E \leq G$.*

Proof. One direction is obvious, so assume $X \neq 0$. Look at the triangle

$$pk \otimes X \to X \to tk \otimes X \to .$$

Either $tk \otimes X \neq 0$ or $pk \otimes X \neq 0$. In the first case we're in $\mathsf{StMod}(kG)$ and we can use Chouinard's theorem for $\mathsf{StMod}(kG)$. In the second case X is not acyclic so its restriction to any subgroup is non-zero. The trivial subgroup is elementary abelian, so we are done. $\square$

5.3.5 The main theorem

Theorem 5.19. *As a tensor triangulated category, $\mathsf{K}(\mathsf{Inj}\, kG)$ is stratified by the canonical action of $H^*(G, k)$.*

Proof. We must prove that if $\mathfrak{p} \in \mathcal{V}_G$, then $\Gamma_{\mathfrak{p}}\mathsf{K}(\mathsf{Inj}\, kG)$ is minimal among tensor ideal localising subcategories of $\mathsf{K}(\mathsf{Inj}\, kG)$.

Let $0 \neq X \in \Gamma_{\mathfrak{p}}\mathsf{K}(\mathsf{Inj}\, kG)$. By Theorem 5.18, there exists E_0 elementary abelian with $X{\downarrow}_{E_0} \neq 0$. Choose $\mathfrak{q}_0 \in \mathcal{V}_{E_0}(X{\downarrow}_{E_0})$. We have

$$\mathrm{res}^*_{G,E_0}(\mathfrak{q}_0) \in \mathcal{V}_G(X{\downarrow}_{E_0}{\uparrow}^G) \subseteq \mathcal{V}_G(X) = \{\mathfrak{p}\},$$

so $\mathrm{res}^*_{G,E_0}(\mathfrak{q}_0) = \mathfrak{p}$.

So there exists $(E, \mathfrak{q})$ with $E \leq E_0$, $\mathfrak{q} \in \mathcal{V}_E$, $\mathrm{res}^*_{E_0,E}\,\mathfrak{q} = \mathfrak{q}_0$, and $\mathfrak{p}$ originates in $\mathfrak{q}$. By the Theorem 5.17 we have $\mathfrak{q} \in \mathcal{V}_E(X{\downarrow}_E)$, i.e., $\Gamma_{\mathfrak{q}}X{\downarrow}_E \neq 0$.

By Theorem 5.13, all $(E, \mathfrak{q})$ where $\mathfrak{p}$ originate in E are conjugate. It follows that if $\Gamma_{\mathfrak{q}}X{\downarrow}_E \neq 0$ for one of these, then the same holds for all of them. So if we choose one, then every $0 \neq X \in \Gamma_{\mathfrak{p}}\mathsf{K}(\mathsf{Inj}\,kG)$ has $\Gamma_{\mathfrak{q}}X{\downarrow}_E \neq 0$, and $X{\downarrow}_E$ is a direct sum of $N_G(E)$-conjugates of this.

Let $0 \neq Y \in \Gamma_{\mathfrak{p}}\mathsf{K}(\mathsf{Inj}\,kG)$ and set $Z = ik_E{\uparrow}^G$, namely the injective resolution of the permutation module on the cosets of E. Then we have

$$
\begin{aligned}
\mathrm{Hom}^*_{kG}(X \otimes_k Z, Y) &= \mathrm{Hom}^*_{kG}(X \otimes_k ik_E{\uparrow}^G, Y) \\
&\cong \mathrm{Hom}^*_{kG}((X{\downarrow}_E \otimes_k ik){\uparrow}^G, Y) \\
&\cong \mathrm{Hom}^*_{kG}(X{\downarrow}_E{\uparrow}^G, Y) \\
&\cong \mathrm{Hom}^*_{kE}(X{\downarrow}_E, Y{\downarrow}_E) \\
&\cong \bigoplus_{\mathfrak{q}} \mathrm{Hom}^*_{kE}(\Gamma_{\mathfrak{q}}X{\downarrow}_E, \Gamma_{\mathfrak{q}}Y{\downarrow}_E).
\end{aligned}
$$

Since both $\Gamma_{\mathfrak{q}}X{\downarrow}_E$ and $\Gamma_{\mathfrak{q}}Y{\downarrow}_E$ are both non-zero, by the classification of localising subcategories of $\mathsf{K}(\mathsf{Inj}\,kE)$ and Lemma 4.4 we have $\mathrm{Hom}^*_{kE}(\Gamma_{\mathfrak{q}}X{\downarrow}_E, \Gamma_{\mathfrak{q}}Y{\downarrow}_E) \neq 0$.

So using the minimality test of Lemma 4.11 for the tensor triangulated category $\mathsf{K}(\mathsf{Inj}\,kG)$, we deduce that $\Gamma_{\mathfrak{p}}\mathsf{K}(\mathsf{Inj}\,kG)$ is minimal among tensor ideal localising subcategories.   □

Using the discussion of Section 3.3, this implies the following result for the stable module category.

Theorem 5.20. *As a tensor triangulated category, $\mathsf{StMod}(kG)$ is stratified by the canonical action of $H^*(G, k)$.*   □

5.4   Exercises

In what follows T denotes a compactly generated R-linear triangulated category. Ideals and elements in R will be assumed to be homogeneous.

(1) Assume T is a tensor triangulated category. Let $\mathfrak{a}$ and $\mathfrak{b}$ be ideals in R. Prove that if $\mathfrak{a} \subseteq \mathfrak{b}$, then $X/\!\!/\mathfrak{b}$ is in $\mathrm{Thick}_{\mathsf{T}}(X/\!\!/\mathfrak{a})$. Deduce that if $\sqrt{\mathfrak{a}} = \sqrt{\mathfrak{b}}$, then

$$
\mathrm{Thick}_{\mathsf{T}}(X/\!\!/\mathfrak{a}) = \mathrm{Thick}_{\mathsf{T}}(X/\!\!/\mathfrak{b}).
$$

(2) Prove that for each ideal $\mathfrak{a}$ of R and object X in T one has

$$
\Gamma_{\mathcal{V}(\mathfrak{a})}X \in \mathrm{Loc}_{\mathsf{T}}(X/\!\!/\mathfrak{a}) \quad \text{and} \quad X/\!\!/\mathfrak{a} \in \mathrm{Thick}_{\mathsf{T}}(\Gamma_{\mathcal{V}(\mathfrak{a})}X).
$$

Here T need not be tensor triangulated.

Hint: Induce on the number of generators of $\mathfrak{a}$. Use the description of $\Gamma_{\mathcal{V}(r)}X$ from Proposition 4.27.

(3) Using the previous exercise prove that for each specialisation closed subset $\mathcal{V}$ of $\operatorname{Spec} R$, and any decomposition $\bigcup_{i \in I} V(\mathfrak{a}_i) = \mathcal{V}$, and set G of compact generators for T, there are equalities

$$\mathsf{T}_{\mathcal{V}} = \operatorname{Loc}_{\mathsf{T}}(\Gamma_{V(\mathfrak{a}_i)} C \mid C \in G) = \operatorname{Loc}_{\mathsf{T}}(C /\!\!/ \mathfrak{a}_i \mid C \in G)$$

This proves that $\mathsf{T}_{\mathcal{V}}$ is compactly generated, so the exact functors $\Gamma_{\mathcal{V}}$ and $L_{\mathcal{V}}$ are smashing, meaning that they commute with coproducts in T.

(4) Let $E = \langle g \mid g^p = 1 \rangle$ and k be a field of characteristic p. Write down a basis for the Koszul construction on kE with respect to $z = g - 1$, namely the differential graded algebra $A = kE\langle y \rangle$ with $|y| = -1$, $y^2 = 0$ and $dy = z$.

Prove that the inclusion of the exterior algebra on the element $z^{p-1}y$ of degree -1 into A is a quasi-isomorphism.

(5) Using tensor products and the Künneth theorem, show that for a general elementary abelian p-group $E = \langle g_1, \ldots, g_r \rangle$, the inclusion of an exterior algebra on the elements $z_i^{p-1} y_i$ $(i \le i \le r)$ into the Koszul construction is a quasi-isomorphism.

(6) Let A be a local ring, with residue field k. Prove that a complex $M \in \mathsf{D}(A)$ is in $\operatorname{Thick}(k)$ if and only if the length of the A-module $H^* M$ is finite.

Hint: For the converse, first consider the case where M is a module; when M is a complex, induce on the number of non-zero cohomology modules of M.

Appendix A. Support for modules over commutative rings

Let A be a commutative noetherian ring. We consider the category $\mathsf{Mod}\,A$ of A-modules and its full subcategory $\mathsf{mod}\,A$ which is formed by all finitely generated A-modules. Note that an A-module is finitely generated if and only if it is noetherian.

The *spectrum* $\operatorname{Spec} A$ of A is the set of prime ideals in it. A subset of $\operatorname{Spec} A$ is *Zariski closed* if it is of the form

$$\mathcal{V}(\mathfrak{a}) = \{\mathfrak{p} \in \operatorname{Spec} A \mid \mathfrak{a} \subseteq \mathfrak{p}\}$$

for some ideal $\mathfrak{a}$ of A. A subset $\mathcal{V}$ of $\operatorname{Spec} A$ is *specialisation closed* if for any pair $\mathfrak{p} \subseteq \mathfrak{q}$ of prime ideals, $\mathfrak{p} \in \mathcal{V}$ implies $\mathfrak{q} \in \mathcal{V}$. The *specialisation closure* of a subset $\mathcal{U} \subseteq \operatorname{Spec} A$ is the subset

$$\operatorname{cl}\mathcal{U} = \{\mathfrak{p} \in \operatorname{Spec} A \mid \text{there exists } \mathfrak{q} \in \mathcal{U} \text{ with } \mathfrak{q} \subseteq \mathfrak{p}\}\,.$$

This is the smallest specialisation closed subset containing $\mathcal{U}$.

A.1   Big support

The *big support* of an A-module M is the subset

$$\operatorname{Supp}_A M = \{\mathfrak{p} \in \operatorname{Spec} A \mid M_{\mathfrak{p}} \neq 0\}\,.$$

Observe that this is a specialisation closed subset of $\operatorname{Spec} A$.

Lemma A.1. *One has* $\operatorname{Supp}_A A/\mathfrak{a} = \mathcal{V}(\mathfrak{a})$ *for each ideal* $\mathfrak{a}$ *of* A.

Proof. Fix $\mathfrak{p} \in \operatorname{Spec} A$ and let $S = A \setminus \mathfrak{p}$. Recall that for any A-module M, an element x/s in $S^{-1}M = M_{\mathfrak{p}}$ is zero iff there exists $t \in S$ such that $tx = 0$. Thus we have $(A/\mathfrak{a})_{\mathfrak{p}} = 0$ iff there exists $t \in S$ with $t(1 + \mathfrak{a}) = t + \mathfrak{a} = 0$ iff $\mathfrak{a} \not\subseteq \mathfrak{p}$. □

Lemma A.2. *If* $0 \to M' \to M \to M'' \to 0$ *is an exact sequence of A-modules, then* $\operatorname{Supp}_A M = \operatorname{Supp}_A M' \cup \operatorname{Supp}_A M''$.

Proof. The sequence $0 \to M'_{\mathfrak{p}} \to M_{\mathfrak{p}} \to M''_{\mathfrak{p}} \to 0$ is exact for each $\mathfrak{p}$ in Spec A. □

Lemma A.3. *Let $M = \sum_i M_i$ be an A-module, written as a sum of submodules M_i. Then $\operatorname{Supp}_A M = \bigcup_i \operatorname{Supp}_A M_i$.*

Proof. The assertion is clear if the sum $\sum_i M_i$ is direct, since

$$\bigoplus_i (M_i)_{\mathfrak{p}} = (\bigoplus_i M_i)_{\mathfrak{p}}.$$

As $M_i \subseteq M$ for all i one gets $\bigcup_i \operatorname{Supp}_A M_i \subseteq \operatorname{Supp}_A M$, from Lemma A.2. On the other hand, $M = \sum_i M_i$ is a factor of $\bigoplus_i M_i$, so $\operatorname{Supp}_A M \subseteq \bigcup_i \operatorname{Supp}_A M_i$. □

We write $\operatorname{ann}_A M$ for the ideal of elements in A that annihilate M.

Lemma A.4. *One has $\operatorname{Supp}_A M \subseteq \mathcal{V}(\operatorname{ann}_A M)$, with equality when M is in $\operatorname{mod} A$.*

Proof. Write $M = \sum_i M_i$ as a sum of cyclic modules $M_i \cong A/\mathfrak{a}_i$. Then

$$\operatorname{Supp}_A M = \bigcup_i \operatorname{Supp}_A M_i = \bigcup_i \mathcal{V}(\mathfrak{a}_i) \subseteq \mathcal{V}(\bigcap_i \mathfrak{a}_i) = \mathcal{V}(\operatorname{ann}_A M),$$

and equality holds if the sum is finite. □

Lemma A.5. *Let $M \neq 0$ be an A-module. If $\mathfrak{p}$ is maximal in the set of ideals which annihilate a non-zero element of M, then $\mathfrak{p}$ is prime.*

Proof. Suppose $0 \neq x \in M$ and $\mathfrak{p}x = 0$. Let $a, b \in A$ with $ab \in \mathfrak{p}$ and $a \notin \mathfrak{p}$. Then $(\mathfrak{p}, b)$ annihilates $ax \neq 0$, so the maximality of $\mathfrak{p}$ implies $b \in \mathfrak{p}$. Thus $\mathfrak{p}$ is prime. □

Lemma A.6. *Let $M \neq 0$ be an A-module. There exists a submodule of M which is isomorphic to $A/\mathfrak{p}$ for some prime ideal $\mathfrak{p}$.*

Proof. The set of ideals annihilating a non-zero element has a maximal element, since A is noetherian. Now apply Lemma A.5. □

Lemma A.7. *For each M in $\operatorname{mod} A$ there exists a finite filtration*

$$0 = M_0 \subseteq M_1 \subseteq \ldots \subseteq M_n = M$$

such that each factor M_i/M_{i-1} is isomorphic to $A/\mathfrak{p}_i$ for some prime ideal $\mathfrak{p}_i$. In that case one has $\operatorname{Supp}_A M = \bigcup_i \mathcal{V}(\mathfrak{p}_i)$.

Proof. Repeated application of Lemma A.6 yields a chain of submodules $0 = M_0 \subseteq M_1 \subseteq M_2 \subseteq \ldots$ of M such that each M_i/M_{i-1} is isomorphic to $A/\mathfrak{p}_i$ for some $\mathfrak{p}_i$. This chain stabilises since M is noetherian, and therefore $\bigcup_i M_i = M$.

The last assertion follows from Lemmas A.2 and A.1. □

A.2 Serre subcategories

A full subcategory C of A-modules is called *Serre subcategory* if for every exact sequence $0 \to M' \to M \to M'' \to 0$ of A-modules, M belongs to C if and only if M' and M'' belong to C. We set

$$\operatorname{Supp}_A \mathsf{C} = \bigcup_{M \in \mathsf{C}} \operatorname{Supp}_A M.$$

Proposition A.8. *The assignment* $\mathsf{C} \mapsto \operatorname{Supp}_A \mathsf{C}$ *induces a bijection between*

(1) *the set of Serre subcategories of* $\mathsf{mod}\, A$, *and*

(2) *the set of specialisation closed subsets of* $\operatorname{Spec} A$.

Its inverse takes $\mathcal{V} \subseteq \operatorname{Spec} A$ *to* $\{M \in \mathsf{mod}\, A \mid \operatorname{Supp} M \subseteq \mathcal{V}\}$.

Proof. Both maps are well defined by Lemmas A.2 and A.4. If $\mathcal{V} \subseteq \operatorname{Spec} A$ is a specialisation closed subset, let $\mathsf{C}_{\mathcal{V}}$ denote the smallest Serre subcategory containing $\{A/\mathfrak{p} \mid \mathfrak{p} \in \mathcal{V}\}$. Then we have $\operatorname{Supp} \mathsf{C}_{\mathcal{V}} = \mathcal{V}$, by Lemmas A.1 and A.2. Now let C be a Serre subcategory of $\mathsf{mod}\, A$. Then

$$\operatorname{Supp} \mathsf{C} = \{\mathfrak{p} \in \operatorname{Spec} A \mid A/\mathfrak{p} \in \mathsf{C}\}$$

by Lemma A.7. It follows that $\mathsf{C} = \mathsf{C}_{\mathcal{V}}$ for each Serre subcategory C, where $\mathcal{V} = \operatorname{Supp} \mathsf{C}$. Thus $\operatorname{Supp} \mathsf{C}_1 = \operatorname{Supp} \mathsf{C}_2$ implies $\mathsf{C}_1 = \mathsf{C}_2$ for each pair $\mathsf{C}_1, \mathsf{C}_2$ of Serre subcategories. $\qquad\square$

Corollary A.9. *Let M and N be in* $\mathsf{mod}\, A$. *Then* $\operatorname{Supp}_A N \subseteq \operatorname{Supp}_A M$ *if and only if N belongs to the smallest Serre subcategory containing M.*

Proof. With C denoting the smallest Serre subcategory containing M, there is an equality $\operatorname{Supp}_A \mathsf{C} = \operatorname{Supp}_A M$ by Lemma A.2. Now apply Proposition A.8. $\quad\square$

A.3 Localising subcategories

A full subcategory C of A-modules is said to be *localising* if it is a Serre subcategory and if for any family of A-modules $M_i \in \mathsf{C}$ the sum $\bigoplus_i M_i$ is in C. The result below is from [31, p. 425].

Corollary A.10. *The assignment* $\mathsf{C} \mapsto \operatorname{Supp}_A \mathsf{C}$ *gives a bijection between*

(1) *the set of localising subcategories of* $\mathsf{Mod}\, A$, *and*

(2) *the set of specialisation closed subsets of* $\operatorname{Spec} A$.

Its inverse takes $\mathcal{V} \subseteq \operatorname{Spec} A$ *to* $\{M \in \mathsf{Mod}\, A \mid \operatorname{Supp}_A M \subseteq \mathcal{V}\}$.

Proof. The proof is essentially the same as the one of Proposition A.8 if we observe that any A-module M is the sum $M = \sum_i M_i$ of its finitely generated submodules. Note that M belongs to a localising subcategory C if and only if all M_i belong to C. In addition, we use that $\operatorname{Supp}_A M = \bigcup_i \operatorname{Supp}_A M_i$; see Lemma A.3. $\quad\square$

A.4   Injective modules

The following proposition collects the basic properties of injective modules over a commutative noetherian ring; for a proof see [43, §18]. For each $\mathfrak{p} \in \operatorname{Spec} A$ we denote $E(A/\mathfrak{p})$ the injective hull of $A/\mathfrak{p}$.

Proposition A.11.   (1) *An arbitrary direct sum of injective modules is injective.*

(2) *Every injective module decomposes essentially uniquely as a direct sum of injective indecomposables.*

(3) *$E(A/\mathfrak{p})$ is indecomposable for each $\mathfrak{p}$ in $\operatorname{Spec} A$.*

(4) *Each injective indecomposable is isomorphic to $E(A/\mathfrak{p})$ for a unique prime ideal $\mathfrak{p}$.*   □

Let $\mathfrak{p}$ a prime ideal in A and let M be an A-module. The module M is said to be $\mathfrak{p}$-*torsion* if each element of M is annihilated by a power of $\mathfrak{p}$; equivalently:

$$M = \{x \in M \mid \text{there exists an integer } n \geq 0 \text{ such that } \mathfrak{p}^n \cdot x = 0\}.$$

The module M is $\mathfrak{p}$-*local* if the natural map $M \to M_{\mathfrak{p}}$ is bijective.

For example, $A/\mathfrak{p}$ is $\mathfrak{p}$-torsion, but it is $\mathfrak{p}$-local only if $\mathfrak{p}$ is a maximal ideal, while $A_{\mathfrak{p}}$ is $\mathfrak{p}$-local, but it is $\mathfrak{p}$-torsion only if $\mathfrak{p}$ is a minimal prime ideal. The A-module $E(A/\mathfrak{p})$ is both $\mathfrak{p}$-torsion and $\mathfrak{p}$-local. Using this observation the following is easy to prove.

Lemma A.12. *Let $\mathfrak{p}$ and $\mathfrak{q}$ be prime ideals in A. Then*

$$E(A/\mathfrak{p})_{\mathfrak{q}} = \begin{cases} E(A/\mathfrak{p}) & \text{if } \mathfrak{q} \in \mathcal{V}(\mathfrak{p}), \\ 0 & \text{otherwise.} \end{cases} \qquad \square$$

A.5   Support

Each A-module M admits a minimal injective resolution

$$0 \to M \to I^0 \to I^1 \to I^2 \to \cdots$$

and such a resolution is unique, up to isomorphism of complexes of A-modules. We say that $\mathfrak{p}$ *occurs* in a minimal injective resolution I of M, if for some integer $i \in \mathbb{Z}$, the module I^i has a direct summand isomorphic to $E(A/\mathfrak{p})$. We call the set

$$\operatorname{supp}_A M = \left\{ \mathfrak{p} \in \operatorname{Spec} A \,\middle|\, \begin{array}{l} \mathfrak{p} \text{ occurs in a minimal} \\ \text{injective resolution of } M \end{array} \right\}$$

the *support of M*. In the literature, it is sometimes referred to as the 'small support' or the 'cohomological support', to distinguish it from the big support $\operatorname{Supp}_A M$.

Lemma A.13. *Let M be an A-module and $\mathfrak{p} \in \operatorname{Spec} A$. If I is a minimal injective resolution of M, then $I_{\mathfrak{p}}$ is a minimal injective resolution of $M_{\mathfrak{p}}$. Therefore*

$$\operatorname{supp}_A(M_{\mathfrak{p}}) = \operatorname{supp}_A M \cap \{\mathfrak{q} \in \operatorname{Spec} A \mid \mathfrak{q} \subseteq \mathfrak{p}\}.$$

Proof. For the first assertion, see for example Lemmas 5 and 6 in [43, §18]. The formula for the support of $M_{\mathfrak{p}}$ then follows from Lemma A.12. □

We write $k(\mathfrak{p})$ for the residue field $A_{\mathfrak{p}}/\mathfrak{p}A_{\mathfrak{p}}$ at $\mathfrak{p} \in \operatorname{Spec} A$.

Lemma A.14. *Let M be an A-module and $\mathfrak{p} \in \operatorname{Spec} A$. The following are equivalent:*

(1) $\mathfrak{p} \in \operatorname{supp}_A M$;

(2) $\operatorname{Ext}^*_{A_{\mathfrak{p}}}(k(\mathfrak{p}), M_{\mathfrak{p}}) \neq 0$;

(3) $\operatorname{Tor}^{A_{\mathfrak{p}}}_*(k(\mathfrak{p}), M_{\mathfrak{p}}) \neq 0$.

Proof. For the equivalence of (1) and (2) see [43, Theorem 18.7]. The equivalence of (2) and (3) is more involved, and was proved by Foxby [30]. □

Lemma A.15. *For each A-module M one has*

$$\operatorname{supp}_A M \subseteq \operatorname{cl}(\operatorname{supp}_A M) = \operatorname{Supp}_A M \subseteq \mathcal{V}(\operatorname{ann} M),$$

and equalities hold when M is finitely generated.

Proof. The equality follows from Lemma A.13, while the inclusions are obvious.

Suppose now M is finitely generated. Given Lemma A.4, to prove that equalities hold, it remains to verify $\operatorname{Supp}_A M \subseteq \operatorname{supp}_A M$. If $M_{\mathfrak{p}} \neq 0$ for some $\mathfrak{p} \in \operatorname{Spec} A$, then $k(\mathfrak{p}) \otimes_{A_{\mathfrak{p}}} M_{\mathfrak{p}} \neq 0$ by Nakayama's Lemma, for $M_{\mathfrak{p}}$ is a finitely generated module over the local ring $A_{\mathfrak{p}}$. In particular, $\operatorname{Tor}^{A_{\mathfrak{p}}}_*(k(\mathfrak{p}), M_{\mathfrak{p}}) \neq 0$, and hence $\mathfrak{p}$ is in $\operatorname{supp}_A M$, by Lemma A.14. □

A.6 Specialization closed sets

Given a subset $\mathcal{U} \subseteq \operatorname{Spec} A$, we consider the full subcategory

$$\mathsf{M}_{\mathcal{U}} = \{M \in \operatorname{Mod} A \mid \operatorname{supp}_A M \subseteq \mathcal{U}\}.$$

The next result does not hold for arbitrary subsets of $\operatorname{Spec} A$. In fact, it can be used to characterise the property that $\mathcal{V}$ is specialisation closed.

Lemma A.16. *Let $\mathcal{V}$ be a specialisation closed subset of $\operatorname{Spec} A$. Then for each A-module M, one has*

$$\operatorname{supp}_A M \subseteq \mathcal{V} \iff M_{\mathfrak{q}} = 0 \text{ for each } \mathfrak{q} \text{ in } \operatorname{Spec} A \setminus \mathcal{V}.$$

The subcategory $\mathsf{M}_{\mathcal{V}}$ of $\operatorname{Mod} A$ is closed under set-indexed direct sums, and in any exact sequence $0 \to M' \to M \to M'' \to 0$ of A-modules, M is in $\mathsf{M}_{\mathcal{V}}$ if and only if M' and M'' are in $\mathsf{M}_{\mathcal{V}}$.

Proof. Since $\mathcal{V}$ is specialisation closed, it contains $\mathrm{supp}_A M$ if and only if it contains $\mathrm{cl}(\mathrm{supp}_A M)$. Thus the first statement is a consequence of Lemma A.15. Given this the second statement follows, since for each $\mathfrak{q}$ in $\mathrm{Spec}\, A$, the functor taking an A-module M to $M_{\mathfrak{q}}$ is exact and preserves set-indexed direct sums. □

Torsion modules and local modules can be recognised from their supports.

Lemma A.17. *Let M be an A-module and $\mathfrak{p} \in \mathrm{Spec}\, A$. Then*

(1) *M is $\mathfrak{p}$-local if and only if $\mathrm{supp}_A M \subseteq \{\mathfrak{q} \in \mathrm{Spec}\, A \mid \mathfrak{q} \subseteq \mathfrak{p}\}$, and*

(2) *M is $\mathfrak{p}$-torsion if and only if $\mathrm{supp}_A M \subseteq \mathcal{V}(\mathfrak{p})$.*

Proof. Let I be a minimal injective resolution of M.

(1) Since $I_{\mathfrak{p}}$ is a minimal injective resolution of $M_{\mathfrak{p}}$, and minimal injective resolutions are unique up to isomorphism, $M \xrightarrow{\sim} M_{\mathfrak{p}}$ if and only if $I \xrightarrow{\sim} I_{\mathfrak{p}}$. This implies the desired equivalence, by Lemma A.12.

(2) When $\mathrm{supp}_A M \subseteq \mathcal{V}(\mathfrak{p})$, then, by definition of support, one has that I^0 is isomorphic to a direct sum of copies of $E(A/\mathfrak{q})$ with $\mathfrak{q} \in \mathcal{V}(\mathfrak{p})$. Since each $E(A/\mathfrak{q})$ is $\mathfrak{p}$-torsion, so is I^0, and hence the same is true of M, for it is a submodule of I^0.

Conversely, when M is $\mathfrak{p}$-torsion, $M_{\mathfrak{q}} = 0$ for each $\mathfrak{q}$ in $\mathrm{Spec}\, A$ with $\mathfrak{q} \not\supseteq \mathfrak{p}$. This implies $\mathrm{supp}_A M \subseteq \mathcal{V}(\mathfrak{p})$, by Lemma A.16. □

Lemma A.18. *Let $\mathfrak{p}$ be a prime ideal in A and set $\mathcal{U} = \{\mathfrak{q} \in \mathrm{Spec}\, A \mid \mathfrak{q} \subseteq \mathfrak{p}\}$.*

(1) *Restriction along the morphism $A \to A_{\mathfrak{p}}$ identifies $\mathsf{Mod}\, A_{\mathfrak{p}}$ with the subcategory $\mathsf{M}_{\mathcal{U}}$ of $\mathsf{Mod}\, A$. Therefore $\mathsf{M}_{\mathcal{U}}$ is closed under taking kernels, cokernels, extensions, direct sums, and products.*

(2) *If M, N are A-modules and one of them belongs to $\mathsf{M}_{\mathcal{U}}$, then $\mathrm{Hom}_A(M, N)$ is in $\mathsf{M}_{\mathcal{U}}$.*

Proof. (1) The objects in the subcategory $\mathsf{M}_{\mathcal{U}}$ are precisely the $\mathfrak{p}$-local A-modules, by Lemma A.17. Thus the inclusion functor has a left and a right adjoint. It follows that $\mathsf{M}_{\mathcal{U}}$ is an exact abelian and extension closed subcategory of $\mathsf{Mod}\, A$, closed under set-indexed direct sums and products.

(2) The action of A on $\mathrm{Hom}_A(M, N)$ factors via $\mathrm{End}_A(M)$ and $\mathrm{End}_A(N)$. If M or N is $\mathfrak{p}$-local, then this action factors through the map $A \to A_{\mathfrak{p}}$. □

Bibliography

[1] L. L. Avramov and R.-O. Buchweitz, *Support varieties and cohomology over complete intersections*, Invent. Math. **142** (2000), 285–318.

[2] L. L. Avramov, R.-O. Buchweitz, S. B. Iyengar, and C. Miller, *Homology of perfect complexes*, Adv. Math. **223** (2010), 1731–1781. [Corrigendum: Adv. Math. **225** (2010), 3576–3578.]

[3] A. Beilinson, J. Bernstein, and P. Deligne, *Faisceaux pervers*, Astérisque **100**, Soc. Math. France, 1983.

[4] D. J. Benson, *Representations and Cohomology I: Basic representation theory of finite groups and associative algebras*, Cambridge Studies in Advanced Mathematics, vol. 30, Cambridge University Press, 1991, reprinted in paperback, 1998.

[5] D. J. Benson, *Representations and Cohomology II: Cohomology of groups and modules*, Cambridge Studies in Advanced Mathematics, vol. 31, Cambridge University Press, 1991, reprinted in paperback, 1998.

[6] D. J. Benson, *Cohomology of modules in the principal block of a finite group*, New York Journal of Mathematics **1** (1995), 196–205.

[7] D. J. Benson, *The nucleus, and extensions between modules for a finite group*, Representations of Algebras, Proceedings of the Ninth International Conference (Beijing 2000), Beijing Normal University Press, 2002, second volume.

[8] D. J. Benson, J. F. Carlson, and J. Rickard, *Complexity and varieties for infinitely generated modules, II*, Math. Proc. Camb. Phil. Soc. **120** (1996), 597–615.

[9] D. J. Benson, J. F. Carlson, and J. Rickard, *Thick subcategories of the stable module category*, Fundamenta Mathematicae **153** (1997), 59–80.

[10] D. J. Benson, S. B. Iyengar, and H. Krause, *Local cohomology and support for triangulated categories*, Ann. Scient. Éc. Norm. Sup. (4) **41** (2008), 1–47.

[11] D. J. Benson, S. B. Iyengar, and H. Krause, *Stratifying modular representations of finite groups*, Ann. of Math. **175** (2012), to appear.

[12] D. J. Benson, S. B. Iyengar, and H. Krause, *Stratifying triangulated categories*, J. Topology **4** (2011), 641–666.

[13] D. J. Benson, S. B. Iyengar, and H. Krause, *Colocalizing subcategories and cosupport*, J. Reine Angew. Math., to appear.

[14] D. J. Benson and H. Krause, *Complexes of injective kG-modules*, Algebra & Number Theory **2** (2008), 1–30.

[15] D. J. Benson, H. Krause, and S. Schwede, *Realizability of modules over Tate cohomology*, Trans. Amer. Math. Soc. **356** (2004), 3621–3668.

[16] D. J. Benson, H. Krause, and S. Schwede, *Introduction to realizability of modules over Tate cohomology*, Fields Inst. Comm., vol. 45, American Math. Society, 2005, pp. 81–97.

[17] I. N. Bernstein, I. M. Gelfand, S. I. Gelfand, *Algebraic vector bundles on P^n and problems of linear algebra*, Funct. Anal. Appl. **12** (1978), 212–214.

[18] M. Bökstedt and A. Neeman, *Homotopy colimits in triangulated categories*, Compositio Math. **86** (1993), 209–234.

[19] E. H. Brown, Jr., *Cohomology theories*, Ann. of Math. (2) **75** (1962), 467–484.

[20] W. Bruns and J. Herzog, *Cohen-Macaulay rings*, Cambridge Studies in Advanced Mathematics, **39**. Cambridge University Press, Cambridge, 1998. Revised edition.

[21] R.-O. Buchweitz, *Maximal Cohen-Macaulay modules and Tate-cohomology over Gorenstein rings*, preprint, Univ. Hannover 1986; http://hdl.handle.net/1807/16682.

[22] T. Bühler, *Exact categories*, Expo. Math., **28** (2010) 1–69.

[23] J. F. Carlson, *The complexity and varieties of modules*, Integral representations and their applications, Oberwolfach, 1980, Lecture Notes in Mathematics, vol. 882, Springer-Verlag, Berlin/New York, 1981, pp. 415–422.

[24] J. F. Carlson, *The varieties and cohomology ring of a module*, J. Algebra **85** (1983), 104–143.

[25] H. Cartan and S. Eilenberg, *Homological Algebra*, Princeton University Press, Princeton, NJ, 1956.

[26] L. Chouinard, *Projectivity and relative projectivity over group rings*, J. Pure Appl. Algebra **7** (1976), 278–302.

[27] E. C. Dade, *Endo-permutation modules over p-groups, II*, Ann. of Math. **108** (1978), 317–346.

[28] A. D. Elmendorf, I. Kříž, M. A. Mandell, and J. P. May, *Rings, modules and algebras in stable homotopy theory*, Surveys and Monographs, vol. 47, American Math. Society, 1996.

[29] L. Evens, *The cohomology ring of a finite group*, Trans. Amer. Math. Soc. **101** (1961), 224–239.

[30] H.-B. Foxby, *Bounded complexes of flat modules*, J. Pure Appl. Algebra **15** (1979), 149–172.

[31] P. Gabriel, *Des catégories abéliennes*, Bull. Soc. Math. France **90** (1962), 323–448.

[32] P. Gabriel and M. Zisman, *Calculus of fractions and homotopy theory*, Ergebnisse der Mathematik und ihrer Grenzgebiete, vol. 35, Springer-Verlag, New York, 1967.

[33] D. Happel, *Triangulated categories in the representation theory of finite dimensional algebras*, London Math. Soc. Lecture Note Series, vol. 119, Cambridge University Press, 1988.

[34] R. Hartshorne, *Local cohomology: A seminar given by A. Grothendieck (Harvard, 1961)*, Lecture Notes in Math. 41, Springer-Verlag, 1967.

[35] M. J. Hopkins, *Global methods in homotopy theory*, Homotopy Theory, Durham 1985, Lecture Notes in Mathematics, vol. 117, Cambridge University Press, 1987.

[36] M. Hovey, J. H. Palmieri, and N. P. Strickland, *Axiomatic stable homotopy theory*, Mem. AMS, vol. 128, American Math. Society, 1997.

[37] S. Iyengar, *Modules and cohomology over group algebras. One commutative algebraist's perspective*, in: Trends in commutative algebra (Berkeley 2002), Mathematical Sciences Research Inst. Publ. **51**, Cambridge Univ. Press, Cambridge, (2004) 51–86.

[38] S. Iyengar, *The classification of thick subcategories of perfect complexes over commutative noetherian rings*, in Thick subcategories - classifications and applications, Oberwolfach Report No. 8/2006.

[39] S. Iyengar, G. Leuschke, A. Leykin, C. Miller, E. Miller, A. Singh, and U. Walther, *Twenty-four hours of local cohomology*, Graduate Stud. Math. **87**, American Mathematical Society, Providence, RI, 2007.

[40] B. Keller, *Deriving DG categories*, Ann. Scient. Éc. Norm. Sup. (4) **27** (1994), 63–102.

[41] H. Krause, *The stable derived category of a noetherian scheme*, Compositio Math. **141** (2005), 1128–1162.

[42] H. Krause, *Derived categories, resolutions, and Brown representability*, in *Interactions between homotopy theory and algebra*, 101–139, Contemp. Math., 436 Amer. Math. Soc., Providence, 2007.

[43] H. Matsumura, *Commutative ring theory*, Cambridge University Press (1986).

[44] A. Neeman, *The chromatic tower for D(R)*, Topology **31** (1992), 519–532.

[45] A. Neeman, *The connection between the K-theory localization theorem of Thomason, Trobaugh and Yao and the smashing subcategories of Bousfield and Ravenel*, Ann. Sci. École Norm. Sup. (4) **25** (1992), no. 5, 547–566.

[46] A. Neeman, *The Grothendieck duality theorem via Bousfield's techniques and Brown representability*, J. Amer. Math. Soc. **9** (1996), 205–236.

[47] A. Neeman, *Triangulated categories*, Annals of Mathematics Studies 148, Princeton University Press, 2001.

[48] D. G. Quillen, *A cohomological criterion for p-nilpotence*, J. Pure & Applied Algebra **1** (1971), 361–372.

[49] D. G. Quillen, *The spectrum of an equivariant cohomology ring, I*, Ann. of Math. **94** (1971), 549–572.

[50] D. Quillen, *Higher algebraic K-theory. I*, in *Algebraic K-theory, I: Higher K-theories (Proc. Conf., Battelle Memorial Inst., Seattle, Wash., 1972)*, 85–147. Lecture Notes in Math., 341, Springer, Berlin, 1973.

[51] D. C. Ravenel, *Localization with respect to certain periodic homology theories*, Amer. J. of Math., **106** (1984), 351–414.

[52] J. Rickard, *Derived categories and stable equivalence*, J. Pure Appl. Algebra **61** (1989), 303–317.

[53] J. Rickard, *Idempotent modules in the stable category*, J. London Math. Soc. **178** (1997), 149–170.

[54] J.-L. Verdier, *Des catégories dérivées des catégories abéliennes*, Astérisque No. 239 (1996), xii+253 pp. (1997).

[55] C. Weibel, *Homological algebra*, Cambridge Studies in Advanced Mathematics 38, Cambridge University Press, 1994.

Index

acyclic, 35
    complex, 58
adjunction isomorphism, 11
$\mathrm{ann}_A M$, 94
$\mathfrak{a}$-torsion, 48
$\mathfrak{a}$-torsion functor, 80
augmentation, 11
Auslander–Buchsbaum–Serre
    theorem, 29

BG, 38
BGG correspondence, 83
big support, 93
block, 60
Brown representability
    theorem, 33, 66

canonical action, 66
Chouinard's Theorem, 5
classifying space, 38
cohomological variety, 40, 42
cohomology, 21
    elementary abelian group, 14
    quaternions, 39
    $\mathbb{Z}$, 39
coinduction, 12
colocalising subcategory, 35
compact object, 22
compactly generated, 22
complex, 17
    shift, 17
    suspension, 17
$C^{\perp}$, 35
$^{\perp}C$, 35
cup product, 36

cyclic shifted subgroup, 6

Dade's Lemma, 5
$\mathsf{D}^{\mathsf{b}}(\mathrm{mod}\, A)$, 27
derived category, 21, 55, 61, 83
diagonal action, 10
differential graded
    algebra, 60, 82
    module, 60
dihedral group, 2
$\mathsf{D}(A)$, 27
$\mathsf{D}(\mathrm{Mod}\, A)$, 22

EG, 38
elementary abelian p-group, 5, 10
elementary abelian 2-group, 82
enough injectives, 18
enough projectives, 18
Evens' theorem, 37
exact category, 18
    admissible epi, 18
    admissible mono, 18
    exact sequence, 18
    extension-closed subcategory, 18
    kernel-cokener pair, 18

F-isomorphism, 87
finite representation type, 2
finitely built, 28
free action on sphere, 40
free resolution, 39
Frobenius
    reciprocity, 12
    twist, 41

Oberwolfach Seminars

The workshops organized by the *Mathematisches Forschungsinstitut Oberwolfach* are intended to introduce students and young mathematicians to current fields of research. By means of these well-organized seminars, also scientists from other fields will be introduced to new mathematical ideas. The publication of these workshops in the series *Oberwolfach Seminars* (formerly *DMV Seminar*) makes the material available to an even larger audience.

■ **OWS 42: Dörfler, W. / Lechleiter, A. / Plum, M. / Schneider, G. / Wieners, C.**, Photonic Crystals: Mathematical Analysis and Numerical Approximation (2011).
ISBN 978-3-0348-0112-6

This volume collects a series of lectures which provide an introduction to the mathematical background needed for the modeling and simulation of light, in particular in periodic media, and for its applications in optical devices.

The book concentrates on the mathematics of photonic crystals, which form an important class of physical structures investigated in nanotechnology. Photonic crystals are materials which are composed of two or more different dielectrics or metals, and which exhibit a spatially periodic structure, typically at the length scale of hundred nanometers.

In the mathematical analysis and the numerical simulation of the partial differential equations describing nanostructures, several mathematical difficulties arise, e. g., the appropriate treatment of nonlinearities, simultaneous occurrence of continuous and discrete spectrum, multiple scales in space and time, and the ill-posedness of these problems.

■ **OWS 41: Hacon, C.D. / Kovács, S.**, Classification of Higher Dimensional Algebraic Varieties (2010).
ISBN 978-3-0346-0289-1

This book focuses on recent advances in the classification of complex projective varieties. It is divided into two parts. The first part gives a detailed account of recent results in the minimal model program. In particular, it contains a complete proof of the theorems on the existence

of flips, on the existence of minimal models for varieties of log general type and of the finite generation of the canonical ring. The second part is an introduction to the theory of moduli spaces. It includes topics such as representing and moduli functors, Hilbert schemes, the boundedness, local closedness and separatedness of moduli spaces and the boundedness for varieties of general type.

The book is aimed at advanced graduate students and researchers in algebraic geometry..

■ **OWS 40: Baum, H. / Juhl, A.**, Conformal Differential Geometry. Q-Curvature and Conformal Holonomy (2010).
ISBN 978-3-7643-9908-5

■ **OWS 39: Drton, M. / Sturmfels, B. / Sullivant, S.**, Lectures on Algebraic Statistics (2008).
ISBN 978-3-7643-8904-8

■ **OWS 38: Bobenko, A.I. / Schröder, P. / Sullivan, J.M. / Ziegler, G.M. (Eds.)**, Discrete Differential Geometry (2008).
ISBN 978-3-7643-8620-7

■ **OWS 37: Galdi, G.P. / Rannacher, R. / Robertson, A.M. / Turek, S.**, Hemodynamical Flows (2008).
ISBN 978-3-7643-7805-9

■ **OWS 36: Cuntz, J. / Meyer, R. / Rosenberg, J.M.**, Topological and Bivariant K-theory (2007).
ISBN 978-3-7643-8398-5

■ **OWS 35: Itenberg, I. / Mikhalkin, G. / Shustin, E.**, Tropical Algebraic Geometry (2007).
ISBN 978-3-7643-8309-1